# SpringerBriefs in Geography

SpringerBriefs in Geography presents concise summaries of cutting-edge research and practical applications across the fields of physical, environmental and human geography. It publishes compact refereed monographs under the editorial supervision of an international advisory board with the aim to publish 8 to 12 weeks after acceptance. Volumes are compact, 50 to 125 pages, with a clear focus. The series covers a range of content from professional to academic such as: timely reports of state-of-the art analytical techniques, bridges between new research results, snapshots of hot and/or emerging topics, elaborated thesis, literature reviews, and in-depth case studies.

The scope of the series spans the entire field of geography, with a view to significantly advance research. The character of the series is international and multidisciplinary and will include research areas such as: GIS/cartography, remote sensing, geographical education, geospatial analysis, techniques and modeling, landscape/regional and urban planning, economic geography, housing and the built environment, and quantitative geography. Volumes in this series may analyze past, present and/or future trends, as well as their determinants and consequences. Both solicited and unsolicited manuscripts are considered for publication in this series.

SpringerBriefs in Geography will be of interest to a wide range of individuals with interests in physical, environmental and human geography as well as for researchers from allied disciplines.

Pilar Mercader-Moyano · Manuel Ramos Martín

# Sustainable Renovation of Buildings

Building Information Modelling

 Springer

Pilar Mercader-Moyano (iD)
Higher Technical School of Architecture
University of Seville
Sevilla, Spain

Manuel Ramos Martín
Higher Technical School of Architecture
Polytechnic University of Madrid
Madrid, Spain

ISSN 2211-4165         ISSN 2211-4173  (electronic)
SpringerBriefs in Geography
ISBN 978-3-031-15142-2    ISBN 978-3-031-15143-9  (eBook)
https://doi.org/10.1007/978-3-031-15143-9

This Springer imprint is published by the registered company Springer Nature Switzerland AG
The registered company address is: Gewerbestrasse 11, 6330 Cham, Switzerland

# Preface

Climatic emergency and socioeconomic crisis caused by recent events—COVID-19 and hydrocarbons crisis derived from Russia–Ukrainian war—are the two principal struggles we face as society. European politics, embodied by National Energy and Climate Plans (NECPs) developed by each region, give the way to the green transition of different productive sectors.

Our building stock is responsible for approximately 36% of the $CO_2$ emissions in the European Union [1]. For this reason, these policies focus a large part of their efforts on economically incentivizing a new development model for the building sector that is committed to the large-scale renovation of the existing real estate stock and that, through the reduction of energy demand and of emissions, manage to reduce the environmental impact of these.

Next generation EU is the new recovery instrument that aims to mobilize investments towards strategic sectors for the reorientation of the production model that, among other measures, contributes to decarbonization through the promotion of energy efficiency and the deployment of renewable energies [2].

However, it should be noted that these environmental objectives are not circumstantial and have been present in European policies since 2002 [1, 3, 4, 5]. 2020 targets [5], which sought to reduce energy consumption and GHG emissions by 20% and increase renewable energy production by the same proportion by 2020, have been shown to be effective at European level, having exceeded the emission targets by 1% and fulfilled the purpose of producing renewable energy [6]. However, the medium- and long-term strategy includes more ambitious objectives that seek a reduction in net emissions of at least 55% by 2030 compared to 1990 data [6]. This requires additional measures that project a reduction in emissions for the construction sector of approximately 20% compared to the data for the year 2020 in a period of nine years.

Spanish National Energy and Climate Plans (NECPs) define through measure 2.6 the strategy to address this ambitious target [7], which recognizes the promotion of public support programmes as a prior mechanism to renovate urban areas of cities with an important number of dwellings, becoming density and urban dimension additional topics to consider in residential buildings stock renovation. As answer to this NECPs and the Recovery, Transformation and Resilience Plan, it published RD 835/2021, which regulates support programmes to residential renovation and social housing [8].

Despite these plans, the construction market continues promoting mainly new buildings instead of refurbishment buildings, as shown in the national statistical data of building construction. According to number of building licences given, a 68% of this (283.324) were given to construct new buildings. The refurbishment licences given (131.373) were addressed to perform the facades (42%), roof (40%) and foundations (18%) [9].

Most of the existing national stock of residential buildings were constructed without considering the current Basic Document on Energy Saving of the Technical Building Code (CTE-DB-HE) that transposes energy efficiency European Directives. The 91.6% of buildings were constructed before 2006 [10], year of code publication. Through this obsolescence, the average life expectancy of Spanish dwellings is 80 years and the end of lifespan of 30-years-old dwellings has been calculated during the period 2063–2081 [11].

Moreover, the numbers of second-hand mortgages have been increased the last decade, representing an 87.7% in the last 2021 trimester [12]. This reason, the huge stock of second-hand dwellings, its long lifespan expected, its necessary refurbishment and the economic incentives to make it, lets us think in perfectible models of information that makes it possible to generate knowledge through the proposals of intervention and prioritize the maintenance actions.

Urban and building renovation must be a priority in present context, and it is why we need tools to achieve the objective. The present book presents and contextualizes (from a regulatory and recent studies view) a methodology proposal that considered climate change perspective and social criteria on building renovation interventions.

Sevilla, Spain                                                    Pilar Mercader-Moyano
Madrid, Spain                                                    Manuel Ramos Martín

# References

1. European Union. Directive 2018/844/EU of the European Parliament and of the Council of 30 May 2018 amending Directive 2010/31/EU on the energy performance of buildings and Directive 2012/27/EU on energy efficiency. Available online: https://eur-lex.europa.eu/legal-content/EN/TXT/HTML/?uri=CELEX:32018L0844&from=ES. Accessed on 1 May 2022
2. Verwey M, Langedijk S, Kuenzel R (2020) Next generation EU: a recovery plan for Europe. VOX CEPR policy portal. Available online: https://voxeu.org/article/next-generation-eu-recovery-plan-europe. Accessed on 1 May 2022
3. European Union. Directive 2002/91/EC of the European Parliament and of the Council of 16 December 2002 on the energy performance of buildings. Available online: https://eur-lex.europa.eu/legal-content/EN/TXT/HTML/?uri=CELEX:32002L0091&from=ES. Accessed on 1 May 2022

4.  European Union. Directive 2010/31/UE of the European Parliament and of the Council of 19 May 2010 on the energy performance of buildings. Available online: https://eur-lex.europa. eu/legal-content/EN/TXT/HTML/?uri=CELEX:32010L0031&from=ES. Accessed on 1 May 2022
5.  European Union. Directive 2012/27/EU of the European Parliament and of the Council of 25 October 2012 on energy efficiency, amending Directives 2009/125/EC and 2010/30/EU and repealing Directives 2004/8/EC and 2006/32/EC. Available online: https://eur-lex.eur opa.eu/legal-content/EN/TXT/HTML/?uri=CELEX:32012L0027&from=ES. Accessed on 1 May 2022
6.  European Environment Agency (2021) EEA report No 13/2021. Trends and projections in Europe 2021 2021. Available online: https://www.eea.europa.eu/publications/trends-and-pro jections-in-europe-2021/file. Accessed on 1 May 2022
7.  Spain Government (2020) Integrated national energy and climate plan 2021–2030. Available online: https://ec.europa.eu/energy/sites/ener/files/documents/es_final_necp_main_en. pdf. Accessed on 1 May 2022
8.  Spain Government. Royal Decree 853/2021 of 5 October, which regulates the aid programs for residential rehabilitation and social housing under the recovery, transformation and resilience plan. Available online: https://www.boe.es/eli/es/rd/2021/10/05/853. Accessed on 1 May 2022
9.  Spain Government (2022) Building construction statistics. Available online: https://www. mitma.gob.es/informacion-para-el-ciudadano/informacion-estadistica/construccion/constr uccion-de-edificios/construccion-de-edificios-licencias-municipales-de-obra. Accessed on 1 May 2022
10. Spanish National Statistical Institute (2011) Population and housing censuses 2011. Buildings results by autonomous communities and provinces. Buildings used mainly or exclusively for dwellings and number of properties by year of construction. Available online: https://www. ine.es/up/EARteouL. Accessed on 1 May 2022
11. Rincón L, Pérez G, Cabeza LF (2013) Service life of the dwelling stock in Spain. Int J Life Cycle Assess 18(5):919–925
12. Spanish Mortgage Association (2022) Quarterly statistical bulletin (Fourth quarter 2021) Available online: http://www.ahe.es/bocms/sites/ahenew/estadisticas/boletin-estadistico/arc hivos/Boletin-trimestral-AHE-4T2021.pdf

# Contents

**1 Sustainable Renovation of Buildings and Methodologies
to Quantify Environmental and Economic Impact** ................. 1
  1.1   From Local to Global: Scale ............................... 1
      1.1.1   Urban Sustainable Interventions ...................... 1
      1.1.2   Building's Renovation Sustainability ................... 3
  References ............................................... 4

**2 Spanish Regulations and European Standardization: Driving
Transformation** .............................................. 7
  2.1   Spanish Regulations on Energy Retrofitting and Renovation
      of Buildings ............................................. 7
  2.2   Standards on Sustainable Construction ...................... 9

**3 Methods** ................................................... 15
  3.1   Assessment Purpose ...................................... 15
      3.1.1   Purpose ........................................... 15
      3.1.2   Preview Use ....................................... 15
  3.2   Assessment Object Specification ........................... 16
      3.2.1   Reference Study Period (RSP) ....................... 16
      3.2.2   System Limits ...................................... 16
      3.2.3   Building Model. Physical Characteristics ............... 17
  3.3   Scenario's Definition, Indicators Selection and Quantify
      Methods to Assess Comprehensive Sustainability of TERP ....... 18
      3.3.1   Criteria Justification to Develop Scenario
           and Selection of Indicators from an Environmental
           Perspective ........................................ 19
      3.3.2   Criteria Justification to Develop Scenario
           and Selection of Indicators from a Social Perspective ...... 19
      3.3.3   Criteria Justification to Develop Scenario
           and Selection of Indicators from an Economic
           Perspective ........................................ 20

3.3.4   Environmental Dimension: Selection of Scenarios
and Calculation of Environmental Data to Assess
the Performance According to UNE-EN 15978:2012 ......   20
3.3.5   Social Dimension: Selection of Scenarios
and Calculation of Social Data to Assess
the Performance According to UNE-EN
16309+A1:2015 ..................................   23
3.3.6   Economic Dimension: Selection of Scenarios
and Calculation of Economic Data to Assess
the Performance According to UNE-EN 16627
and UNE-EN 15459-1 .............................   32
3.4   Methods Conclusions ......................................   40
References ....................................................   42

4   Method Application ...........................................   43
4.1   Calculations Through Quantitative and Qualitative Models .......   43
4.2   Discussion ..............................................   45
4.2.1   Results from a Social-Environmental Perspective .........   52
4.2.2   Results from a Social-Economic Perspective .............   53
4.2.3   Results from an Environmental-Economic Perspective .....   54
4.2.4   Optimized Thermal Envelope Renovation Project
(O-TERP) ......................................   55
4.3   Method Application Conclusions ...........................   58

Appendix A ......................................................   61

Appendix B ......................................................   83

Appendix C ......................................................   93

Appendix D ......................................................   103

Index ...........................................................   107

# Abbreviations

| | |
|---|---|
| BEDEC | Product construction database developed by ITEC |
| CDW | Construction and demolition waste |
| COINT | Initial cost |
| COMA | Maintenance cost |
| CYPE | Architecture, engineering and construction software |
| DGT | Spanish directorate-general of traffic |
| DIT | Technical suitability document |
| DO | Design option |
| DOC | Design option combination |
| EAA | Equivalent annual annuity |
| EPD | Environmental product declaration |
| EWL | European Waste List |
| FQM | Final quantitative model |
| GC | Global cost |
| IDEA | Spanish Institute for Diversification and Energy Saving |
| IPCC | Intergovernmental Panel on Climate Change |
| IPREM | Public Indicator of Multiple Effect Income |
| IQM | Initial quantitative model |
| ITEC | Technic Institute of Construction |
| LCA | Life cycle analysis |
| LCC | Life cycle cost |
| NAI | Not assessed indicator |
| NAM | Not assessed module |
| NPV | Net present value |
| OMIE | Iberian electric market operator |
| PGOU | General Urban Development Plan |
| PVPC | Voluntary prices for small consumers |
| REE | Spanish electric net |
| RITE | Spanish Regulation on Thermal Installations in Buildings |

| TERP | Thermal Envelope Regeneration Project |
| TPB€ | Economic time payback |
| VAT | Valued-added tax |

# Chapter 1
# Sustainable Renovation of Buildings and Methodologies to Quantify Environmental and Economic Impact

**Abstract** Sustainability implies to understand our development as society far away from the momentary, the opportune and the superficial, considering the main idea that humans occupy a decentered position, instead of an empowered from natural resources position. IPCC partials reports belong to the sixth cycle of assessment that will publish on 2022 September alerts about the necessity of adopting actions immediately to ensure a livable future, pointing to cities as hotspots of impacts and risks, but a crucial part of the solution AR6 Synthesis Report: Climate Change. Even a narrowing window for action, urbanism, architecture, and construction—as part of a strategy sector responsible of 40% of total GHG emissions to the atmosphere—develop a key role between humanity and nature. Hence, environmental impact reduction of these activities must be a priority to rebalance our ecosystems. We explain in this chapter the objective of different strategies and scales; methods to quantify environmental and economic impacts, and we explore the opportunities of building information modelling to support different way to assessing sustainability.

**Keywords** Urban renovation · Sustainable construction · BIM · Energy retrofitting

## 1.1 From Local to Global: Scale

### 1.1.1 Urban Sustainable Interventions

Urban interventions are multiples, and we need to categorize to distinguish implications. Cervero-Sánchez and Hernández-Aja propose three types: urban remodelling, transformation and renovation [1].

Urban remodelling is an urban reordination with new buildings and public spaces. Urban transformation in a hybrid intervention implies redefinition of public spaces. Finally, urban renovation is a complete intervention that transforms buildings and public spaces to adapt them to actual requirements.

Each intervention depends on a previous diagnostic to detect necessities. The definition of this diagnostic is a complex line research that has as common denominator: the use of urban indicators to decide the typology of intervention.

© The Author(s), under exclusive license to Springer Nature Switzerland AG 2022
P. Mercader-Moyano and M. Ramos Martín, *Sustainable Renovation of Buildings*,
SpringerBriefs in Geography, https://doi.org/10.1007/978-3-031-15143-9_1

The users' perception about the effectiveness of implemented solutions on neighbourhood is fundamental for success of interventions, it is why different strategies to know these perceptions have been researched. Serrano-Jiménez et al. propose an interdisciplinary decision support system to retrofit buildings considering residents' perceptions besides the thermal demands [2]. Mercader-Moyano et al. suggest that COVID-19 has caused the reduction of urban renovation process, and they develop an index methodology towards feasible planning and policymaking under a crisis context [3].

These perceptions are related to a main aspect to get the objective of urban renovation: public and private financial coresponsibility. The integration of neighbour's desires/demands of urban transformations through their participation in the administrative tools is the most effective mechanism to renovate our cities.

From the administrative tools perspective, a twenty representative cases of urban renovation study—which include urban planning, urban design and local environment, buildings and socio-economical dimension—based on Areas of Integrated Renovation (ARI) in cities with over 50,000 inhabitants, determine the evolution of Spanish public policies the last 34 years (until 2012); concluding that a long-time perspective of the intervention that understands the neighbourhood's renovation as a succession of local transformation has been the most effective way to undertake the urban renovation [4].

On the other hand, from the participation perspective, Local Agenda 21 has been seen by various authors [5–9] as the powerfullest tool to drive a sustainable development in our cities. Nevertheless, its voluntary character and the free interpretation of local government administrations of the first Leipzig Charter principles in terms of urban renovation [10] have caused that many of the Local Agendas developed in Spanish context have not dealt directly with this item, it been impulse documents not very proactive and mainly statistical.

An analysis about possibilities of Local Agenda 21 and ecological urban restructuring [11] shows twelve European Model Projects in Leipzig, classified into three typologies of intervention depending on the area: Area I, Urban Ecology, Area II, Rural Development and Area III, Economy and Employment. Authors conclude that meditation instruments in an interdisciplinary process of planning play a main role in the project's success.

Supporting these ideas, Seve et al. [12] propose the creation of a participation methodology that mixed traditional and new tools and classifies them into a taxonomy which is based on approximately forty experiences and using new technologies to agile communications. They categorize the taxonomy into four: D, basic tools; E, time needed; F, typology of urban space and G, purpose. The main objective of this methodology is to impulse a bottom-up mechanism that contributes to the strengthening of the community, which is fundamental whether we must ward off the urban renovation processes from the undesired effects as gentrification. We need to work with local inhabitants and stakeholders, give them voice and let them understand public space as an extension of their homes.

According to Palomeque [13], urban renovation experiences point to (1) the necessity of a common objective of inhabitants to transform their neighbourhoods,

(2) the leadership of local administration into these processes and (3) the concurrency of public–private interests through financial mixed structures. The main objective in this field is to establish criteria—technical, management, procedure and decision-making—to define strategies by stakeholders.

## 1.1.2   Building's Renovation Sustainability

One of the main strategies to face the climate change proposed by European Union along the different directives and supported by next generation funds is energy efficiency, which is also a resilience tool in a current Russia–Ukrainian war and its consequences about energy increase costs.

Energy efficiency tells us about the responsible use of energy resources in every economy sectors. Residential buildings are one of the most important consumption sectors, being responsible of 36% of GHG emissions [14]. It is the main reason why policies and strategies focus on transforming the current building stock by means of two principal measures: (1) decrease energy-demand reduction and (2) increase clean-energy production.

The way to get both measures is different, and let us distinguish between active and passive measures, which is fundamental to order the actions and programme the interventions on buildings. Passive measures focus on the envelope of building, decreasing its energy demand by means of adding isolate to walls/roofs/floors, changing windows/glasses, placing on façade opening blinds/awnings, etc. On the contrary, passive actions do not reduce energy demand, but satisfy energy necessities increasing the performance of heating, ventilation and air conditioning (HVAC) and sanitary hot water (SHW) systems, using renewable fonts and technology.

> **Important.** We focus on passive actions, as the first step to get an energy efficiency of buildings, under a simple principle: **the less energy we need, the less energy we will have to produce**.

Passive actions depend on climate conditions and its cause that we do not give the same answer to different situations, which is regulated by institutions, defining consumption targets to get a nearly zero-energy building (NZEB) by climate regions [15]; quantifying this consumption by means of kWh/m$^2$ per year indicator. Spain—even with its particularities—such as a mainly temperate climate, benefits from the implementation of passive strategies [16–28].

Various research points to passive strategies on temperate climate are an extraordinary effective measure, due to important reduction of isolate thickness in comparison with Centro European climate regions and the unnecessary use of mechanic ventilation with heat recovery [23]. Moreover, it is demonstrated that excessive isolate on thermal envelope causes overheating inside the buildings, which will be increased by climate change current perspectives [21].

Despite the positive effect from energy efficiency perspective of passive actions, every building construction process and the material obtained implies environmental impact that we must consider if we ambitious an urban renovation committed with a comprehensive sustainability view.

For the purpose to determine the environmental impacts of material productions, life cycle analysis (LCA) methodology becomes the most effective way to determine the production impacts. LCA methodology provides different indicators about the impact of a material production—generally of product stage, modules A1–A3—and it is the basis of calculations to create an environmental product declaration (EPD). Global warming potential contribution of each material—expressed by $CO_2$ equivalent per material mass—has been demonstrated as an effective indicator to communicate the impact of a product [29–34] which is commonly known by term "embodied carbon".

# References

1. Cervero-Sánchez N, Agustín-Hernández L (2015) Remodelación, Transformación y Rehabilitación. Tres formas de intervenir en la Vivienda Social del siglo XX. Inf Constr 67:1–11. https://doi.org/10.3989/ic.14.049
2. Serrano-Jiménez A, Lima ML, Molina-Huelva M, Barrios-Padura Á (2019) Promoting urban regeneration and aging in place: APRAM—an interdisciplinary method to support decision-making in building renovation. Sustain Cities Soc 47:101505
3. Mercader-Moyano P, Morat O, Serrano-Jiménez A (2021) Urban and social vulnerability assessment in the built environment: an interdisciplinary index-methodology towards feasible planning and policy-making under a crisis context. Sustain Cities Soc 73:103082
4. Hernandez Aja A, Matesanz Parellada A, Rodriguez-Suarez I, Garcia Madruga C (2015) Evolution of urban renewal policies in Areas of Integrated Renovation in Spain (1978–2012). Inf Constr 67
5. Hernández Aja A (2003) Informe sobre los indicadores locales de sostenibilidad utilizados por los municipios españoles firmantes de la Carta de Aalborg. Relac Int
6. de Manuel Jerez E, Machucha I (2013) Hacia una rehabilitación energética participativa: el caso de Alcosa. In: Comunicaciones y 1a Bienal de Edificación y Urbanismo Sostenible Greencities & Sostenibilidad: Inteligencia Aplicada a La Sostenibilidad Urbana (Edición), pp 253–272
7. Bermejo Gómez De Segura R (2014) Del desarrollo sostenible según Brundtland a la sostenibilidad como biomimesis. Accessed http://publ.hegoa.efaber.net/assets/pdfs/315/Sostenibilidad_DHL.pdf?1399365095
8. Ameen RFM, Mourshed M, Li H (2015) A critical review of environmental assessment tools for sustainable urban design. Environ Impact Assess Rev 55:110–125. https://doi.org/10.1016/j.eiar.2015.07.006
9. Tojo JF, Naredo JM (2010) Libro blanco de la sostenibilidad en el planeamiento urbanístico Español

10. Charter L (2020) The new Leipzig Charter—the transformative power of cities for the common good. In: Ministerial meeting on urban development and territorial cohesion. Available online: https://ec.europa.eu/regional_policy/en/information/publications/brochures/2020/new-leipzig-charter-the-transformative-power-of-cities-for-the-common-good. Accessed 1 May 2022

11. Hahn E, LaFond M (1997) Local Agenda 21 and ecological urban restructuring. A European model project (Berlin). Available online: https://www.econstor.eu/obitstream/10419/49558/1/248079638.pdf. Accessed 1 May 2022

12. Seve B, Redondo E, Sega R (2022) A taxonomy of bottom-up, community planning and participatory tools in the urban planning context. ACE: Arquitectura, Ciudad y Entorno

13. Palomeque GR (2015) Gestión de la rehabilitación sostenible en Grandes Conjuntos de las periferias urbanas por las administraciones públicas locales. Inf Constr 67:1–13

14. European Union. Directive 2018/844/EU of the European Parliament and of the Council of 30 May 2018 amending Directive 2010/31/EU on the energy performance of buildings and Directive 2012/27/EU on energy efficiency. Available online: https://eur-lex.europa.eu/legal-content/EN/TXT/HTML/?uri=CELEX:32018L0844f&from=ES. Accessed 1 May 2022

15. Fosas D, Coley DA, Natarajan S, Herrera M, Fosas de Pando M, Ramallo-Gonzalez A (2018) Mitigation versus adaptation: does insulating dwellings increase overheating risk? Build Environ 143:740–759. https://doi.org/10.1016/j.buildenv.2018.07.033

16. Monge-Barrio A, Sánchez-Ostiz Gutiérrez A (2018a) Residential architecture in Mediterranean climates. Towards optimized passive solutions for the whole year. In: Green energy and technology. Springer Verlag, pp 45–58. https://doi.org/10.1007/978-3-319-69883-0_4

17. Monge-Barrio A, Sánchez-Ostiz A (2018) Passive energy strategies for Mediterranean residential buildings: facing the challenges of climate change and vulnerable populations. Accessed http://search.ebscohost.com/login.aspx?direct=true&db=edsebk&AN=1709244&site=eds-live

18. Monge-Barrio A, Sánchez-Ostiz Gutiérrez A (2018b) Passive energy strategies for Mediterranean residential buildings. Springer International Publishing. https://doi.org/10.1007/978-3-319-69883-0

19. Rodrigues E, Fernandes MS, Gaspar AR, Gomes Á, Costa JJ (2019) Thermal transmittance effect on energy consumption of Mediterranean buildings with different thermal mass. Appl Energy 252:113437. https://doi.org/10.1016/j.apenergy.2019.113437

20. Oregi X, Hernandez P, Hernandez R (2017) Analysis of life-cycle boundaries for environmental and economic assessment of building energy refurbishment projects. Energy Build 136:12–25. https://doi.org/10.1016/J.ENBUILD.2016.11.057

21. Consoli A, Costanzo V, Evola G, Marletta L (2015) Refurbishing an existing apartment block in Mediterranean climate: towards the Passivhaus standard. Energy Procedia 1–10. https://doi.org/10.1016/j.egypro.2017.03.201

22. López-Ochoa LM et al (2018) Environmental and energy impact of the EPBD in residential buildings in hot and temperate Mediterranean zones: the case of Spain. Energy 161:618–634

23. Rodrigues E, Fernandes MS (2020) Overheating risk in Mediterranean residential buildings: comparison of current and future climate scenarios. Appl Energy 259. https://doi.org/10.1016/j.apenergy.2019.114110

24. Blàzquez T, Suàrez R, Ferrari S, Sendra JJ (2021) Addressing the potential for improvement of urban building stock: a protocol applied to a Mediterranean Spanish case. Sustain Cities Soc 71:102967

25. Monzón M, López-Mesa B (2018) Buildings performance indicators to prioritise multi-family housing renovations. Sustain Cities Soc 38:109–122

26. Almeida M, Barbosa R, Malheiro R (2020) Effect of embodied energy on cost-effectiveness of a prefabricated modular solution on renovation scenarios in social housing in Porto, Portugal. Sustainability 12(4):1631

27. Andrade J, Bragança L (2016) Sustainability assessment of dwellings—a comparison of methodologies. Civ Eng Environ Syst 33(2):125–146. https://doi.org/10.1080/10286608.2016.1145676

28. Adalberth K, Almgren A (2001) Life cycle assessment of four multi-family buildings. Int J Low Energy Sustain Build 2:1–21. Accessed https://www.osti.gov/etdeweb/biblio/20195422
29. Dixit MK, Fernández-Solís JL, Lavy S, Culp CH (2012) Need for an embodied energy measurement protocol for buildings: a review paper. Renew Sustain Energy Rev 16(6):3730–3743
30. Dixit MK, Fernández-Solís JL, Lavy S, Culp CH (2010) Identification of parameters for embodied energy measurement: a literature review. Energy Build 42(8):1238–1247
31. Almeida M, Ferreira M, Barbosa R (2018) Relevance of embodied energy and carbon emissions on assessing cost effectiveness in building renovation—contribution from the analysis of case studies in six European countries. Buildings 8(8):103
32. Brás A (2016) Embodied carbon minimisation of retrofit solutions for walls. Proc Inst Civ Eng Eng Sustain 170(3):141–156
33. Brown NW, Olsson S, Malmqvist T (2014) Embodied greenhouse gas emissions from refurbishment of residential building stock to achieve a 50% operational energy reduction. Build Environ 79:46–56
34. International Organization for Standardization (ISO). Sustainability in building construction—general principles (ISO 15392:2008). ISO, Geneva, Switzerland

# Chapter 2
# Spanish Regulations and European Standardization: Driving Transformation

**Abstract** We must distinguish differences between standardization and regulations. Standardization responds to the objective of drawing up technical specifications (standards) that are used on a voluntary basis and with the aim of demonstrating the safety and reliability of certain products and activities—according to AENOR definition; regulations provide legal rules that regulate this activity. This in no way implies that a standard cannot be made legally binding, since means such as legislating in accordance with these standards or referring to them through existing legal provisions are enabled. In the following lines, we will compile the mandatory norms for building regeneration and the voluntary standards, highlighting the potential of renovation and of the comprehensive management of building renovation construction processes.

**Keywords** Sustainability regulations · Sustainable standards · Future perspectives

## 2.1 Spanish Regulations on Energy Retrofitting and Renovation of Buildings

Nowadays, we design our buildings according to European Directives (Energy Performance of Buildings Directive, 2010/31/EU and Energy Efficiency Directive, 2012/27/EU). Each country develops its owns legal frame to address the needed greenhouse emissions objectives through an accurate design adapted to climates zones that lead to reduce consumption of buildings. It means that every city of each country has a climate code to determine transmittance envelope characteristics.

Due to huge Spanish building stock, renovation of buildings is the main strategy to face the needed green transition.

Even of normative restrictions, architects and engineers understand the passive strategies as a secondary factor in buildings to satisfy minimum exigencies of consumption demand, assuming choosing efficient active strategies will satisfy the heating and cooling needs. Moreover, Certification of European energy efficiency promotes this vision due to the main indicators used energy efficiency rating ($kWh/m^2$ year) and environmental impact rating ($kg\, CO_2$ equiv/$m^2$ year), both highly influent by consumption of building.

© The Author(s), under exclusive license to Springer Nature Switzerland AG 2022
P. Mercader-Moyano and M. Ramos Martín, *Sustainable Renovation of Buildings*, SpringerBriefs in Geography, https://doi.org/10.1007/978-3-031-15143-9_2

Despite this, one thing is clear: the most efficient solution is the less-demanding energy according to climate conditions and is directly related to passive strategies.

Normative climate zones are determined according to a static view of climate, but climate is changing, and by only adapting our buildings to new climate scenarios through passive strategies, we can get human comfort regardless of energy fonts.

Many of these passive strategies are inherent to constructive know-how of each culture and additional characteristics of material. The possibilities that it offers to promote low environmental impact interventions adapted to different climate zones are innumerable and needy to explore.

Even the optimistic scenario temperature is increasing, and the normative climate zones and transmittance values of thermal envelope will be obsolete probably before that we can imagine: we need to readapt our goals:

1. Less and Better: Low-Carbon Retrofit
   We need to build less new buildings and retrofit the existing building stock using better solutions. Better solutions imply to use low embodied carbon and energy materials, to promote local constructive local systems and to explore new material with additional characteristics (photo-catalytic materials, phase change material, recycled materials, etc.).
2. Climate Change Design: Resilience
   We must design buildings solutions to considering climate change perspective, thinking thermal envelopes to prevent overheating in our buildings and limiting the dependence of energy primary fonts.
3. Carbon Budget: New Policies
   Building less and better considering climate change perspective is our main purpose, and it is measurable.

As technics, we need to quantify and determine the accuracy of each solution. Economy centres mainly our attention, but we need to quantify our environmental cost by using of a carbon budget. New policies must put effort to promote the use of environmental footprint and industry must be to develop embodied carbon/energy material databases addressing to this objective.

**Important.** How normative and regulations support and structure these necessities?

Limiting the energy consumption of buildings has been a regulatory issue in Spain since 1975, after the 1973 oil crisis. The interest in limiting the consumption of buildings then had a strategic component at an economic level rather than environmental concerns. Since then, a series of measures have been introduced to regulate the reduction of demand in buildings. At the same time, the state housing plans that are being developed are gradually incorporating strategies that are more oriented towards the renovation of buildings than towards the construction of new buildings. However, it was not until 2015, coinciding with the Paris Agreement that measures were more clearly oriented towards the necessary reduction of the environmental impact of buildings, due to the serious risk involved in the—still current—development of our regions.

Table 2.1 summarizes the history of urban renovation and housing regulations, and we identify the range and regulation matter.

At the present time, Royal Decree 835/2021 is one of the greatest supports for the renovation of buildings and cities. The aspects included in the regulation are important from the point of view of how interventions are planned and programmed and allow us to see that the process of defining them is becoming more complex due to the multitude of agents involved.

In this sense, it is worth highlighting the importance of the Programmed Rehabilitation Residential Environments (PRRA), which have specific measures for their renovation and which aim to intervene in them in a comprehensive manner. PRRAs are mainly residential and territorially delimited areas, which opens the door to large-scale neighbourhood interventions that would be financially supported by the administrations.

Interventions on this scale involve more complex management processes, which is why the regulations include the figure of rehabilitation agents and managers, whose objective is to facilitate "turnkey" intervention models.

Royal Decree 835/2021 does not explicitly mention the use of any type of IT tools for information management. However, the multiplicity of aspects to be considered to be eligible for the measures—such as the definition of waste management plans and an accurate quantification of the costs of the intervention—and the participation of users in the decision-making process highlight the importance of the use of building information modelling tools, as they allow for greater control and anticipation of the intervention.

## 2.2  Standards on Sustainable Construction

On the other hand, when looking at the question of standards, we notice that they are more clearly oriented towards the sustainability of construction processes, as well as the sustainability of the energy performance of buildings. Making our way of building increasingly sustainable is an ambitious objective to be achieved, which is why considering work methodologies that favour the use of materials with a low environmental impact and promote the local economy will result in greater sustainability of actions.

Table 2.2 shows a historical review of the different standards that allow project managers to have methodologies that seek greater social, environmental and economic sustainability. We distinguish in the table the dimensions on which they focus within the framework of comprehensive sustainability.

**Table 2.1** Historical list of urban renovation and housing regulations

| Year | Code | Objective | Range | Regulation matter |
|------|------|-----------|-------|-------------------|
| 1975 | Decree 1490/1975 | To establish measures to be adopted in buildings in order to reduce energy consumption | National | Energy efficiency |
| 1979 | Royal Decree 2429/1979 | To approve the basic building standard NBE-CT-79, on thermal conditions in buildings | National | Energy efficiency |
| 1980 | Royal Decree 1618/1980 | To approve the Regulation on Heating, Air Conditioning and Domestic Hot Water Installations, in order to rationalize their energy consumption | National | Energy efficiency |
| 1991 | Royal Decree 1932/1991 | On measures for the financing of eligible housing actions under the 1992–1995 plan | National | Urban renovation |
| 1993 | Directive 93/76/CEE | To limit carbon dioxide emissions by improving energy efficiency (save) | European Union | Energy efficiency |
| 1993 | Royal Decree 726/1993 | To regulate the financing of actions eligible for protection in the rehabilitation of buildings and amending certain articles of Royal Decree 1932/1991 | National | Urban renovation |
| 1995 | Royal Decree 2190/1995 | On measures for the financing of protectable actions in the field of housing and land for the period 1996–1999 | National | Urban renovation |
| 1998 | Royal Decree 1751/1998 | To approve the Regulation on Thermal Installations in Buildings (RITE) and its Complementary Technical Instructions (ITE) and create the Advisory Commission for Thermal Installations in Buildings | National | Energy efficiency |

(continued)

**Table 2.1** (continued)

| Year | Code | Objective | Range | Regulation matter |
|------|------|-----------|-------|-------------------|
| 1998 | Royal Decree 1186/1998 | On measures for the financing of protected housing and land actions under the 1998–2001 plan | National | Urban renovation |
| 2002 | Directive 2002/91/EC | On the energy performance of buildings | European Union | Energy efficiency |
| 2002 | Royal Decree 1/2002 | On financing measures for protected housing and land actions under the 2002–2005 plan | National | Urban renovation |
| 2005 | Royal Decree 801/2005 | To approve the 2005–2008 State Plan to promote citizens' access to housing | National | Urban renovation |
| 2006 | Royal Decree 314/2006 | To approve the technical building code | National | Energy efficiency |
| 2007 | Royal Decree 47/2007 | To approve the basic procedure for the energy performance certification of newly constructed buildings | National | Energy efficiency |
| 2007 | Directive 2010/31/EU | On the energy performance of buildings | European Union | Energy efficiency |
| 2008 | Royal Decree 14/2008 | To modify Royal Decree 801/2005, of 1 July, approving the 2005–2008 State Plan to promote citizens' access to housing | National | Urban renovation |
| 2008 | Royal Decree 2066/2008 | To regulate the State Housing and Rehabilitation Plan 2009–2012 | National | Urban renovation |
| 2010 | Directive 2012/27/EU | On energy efficiency, amending Directives 2009/125/EC and 2010/30/EU and repealing Directives 2004/8/EC and 2006/32/EC text with EEA relevance | European Union | Energy efficiency |

(continued)

**Table 2.1** (continued)

| Year | Code | Objective | Range | Regulation matter |
|---|---|---|---|---|
| 2010 | Royal Decree 1713/2010 | To modify Royal Decree 2066/2008, of 12 December, which regulates the State Housing and Rehabilitation Plan 2009–2012 | | |
| 2013 | Royal Decree 235/2013 | To approve the basic procedure for the certification of the energy performance of buildings | National | Energy efficiency |
| 2015 | Royal Decree Law 7/2015 | To approve the revised text of the Law on Land and Urban Rehabilitation | National | Urban renovation |
| 2016 | Royal Decree 56/2016 | To transpose Directive 2012/27/EU of the European Parliament and of the Council of 25 October 2012 on energy efficiency as regards energy audits, the accreditation of energy service providers and auditors and the promotion of energy supply efficiency | National | Energy efficiency |
| 2018 | Royal Decree 106/2018 | To regulate the State Housing Plan 2018–2021 | National | Urban renovation |
| 2018 | Directive 2018/844/EU | Amending Directive 2010/31/EU on the energy performance of buildings and Directive 2012/27/EU on energy efficiency (text with EEA relevance) | European Union | Energy efficiency |
| 2019 | Housing Plan 2018–2021 | To modify the technical building code, approved by Royal Decree 314/2006 | National | Energy efficiency |

(continued)

**Table 2.1**  (continued)

| Year | Code | Objective | Range | Regulation matter |
|---|---|---|---|---|
| 2021 | Royal Decree Law 19/2021 | On urgent measures to boost building renovation activity in the context of the Recovery, Transformation and Resilience Plan | National | Urban renovation |

**Table 2.2**  Historical list of sustainability on buildings standards

| Year | Code | Objective | Sustainability dimension | Status |
|---|---|---|---|---|
| 2002 | UNE-EN ISO 14020:2002 | Environmental labels and declarations—general principles | Environment | Current |
| 2006 | UNE-EN ISO 14040:2006 | Environmental management—life cycle assessment—principles and framework | Environment | Current |
| 2006 | UNE-EN ISO 14044:2006 | Environmental management—life cycle assessment—requirements and guidelines (ISO 14044:2006) | Environment | Current |
| 2010 | UNE-EN ISO 14025:2010 | Environmental labels and declarations—Type III environmental declarations—principles and procedures (ISO 14025:2006) | Environment | Current |
| 2011 | UNE-CEN/TR 15941:2011 IN | Sustainability of construction works—environmental product declarations—methodology for selection and use of generic data | Environment | Current |
| 2012 | UNE-EN 15978:2012 | Sustainability of construction works—assessment of environmental performance of buildings—calculation method | Environment | Annulled by UNE-EN 15643:2021 |
| 2012 | UNE-EN 15643-1:2012 | Sustainability of construction works—sustainability assessment of buildings—Part 1: general framework | Comprehensive sustainability | Annulled by UNE-EN 15643:2021 |
| 2012 | UNE-EN 15643-2:2012 | Sustainability of construction works—assessment of buildings—Part 2: framework for the assessment of environmental performance | Environment | Annulled by UNE-EN 15643:2021 |

(continued)

**Table 2.2** (continued)

| Year | Code | Objective | Sustainability dimension | Status |
|------|------|-----------|--------------------------|--------|
| 2012 | UNE-EN 15643-3:2012 | Sustainability of construction works—assessment of buildings—Part 3: framework for the assessment of social performance | Social | Annulled by UNE-EN 15643:2021 |
| 2012 | UNE-EN 15643-4:2012 | Sustainability of construction works—assessment of buildings—Part 4: framework for the assessment of economic performance | Economic | Annulled by UNE-EN 15643:2021 |
| 2013 | UNE-EN 15942:2013 | Sustainability of construction works—environmental product declarations—communication format business-to-business | Environment | Annulled by UNE-EN 15942:2022 |
| 2014 | UNE-EN 15804:2012+A1:2014 | Sustainability of construction works—environmental product declarations—core rules for the product category of construction products | Environment | Current |
| 2015 | UNE-EN 16309+A1:2015 | Sustainability of construction works—assessment of social performance of buildings—calculation methodology | Social | Current |
| 2016 | UNE-EN 16627:2016 | Sustainability of construction works—assessment of economic performance of buildings—calculation methods | Economic | Current |
| 2018 | UNE-EN 15459-1:2018 | Energy performance of buildings—economic evaluation procedure for energy systems in buildings—Part 1: calculation procedures, module M1–14 | Economic | Current |
| 2021 | UNE-EN 15643:2021 | Sustainability of construction works—framework for assessment of buildings and civil engineering works | Comprehensive sustainability | Current |
| 2022 | UNE-EN 15942:2022 | Sustainability of construction works—environmental product declarations—communication format business-to-business | Environment | Current |

# Chapter 3
# Methods

**Abstract** A Comprehensive Sustainable Assessment of Thermal Envelope Renovation Project (TERP) and building design options (DO) share a frame method. We need to define system's limits for each scenario depending on the sustainable item to assess. We explain in the following lines a proposal method to integrate sustainable building principles from different standards into building energy retrofitting processes through quantitative and qualitative models that are integrated into a BIM interface.

**Keywords** Sustainable calculation method · Quantification model · Qualification model

## 3.1 Assessment Purpose

### 3.1.1 Purpose

Assessment objective is to quantify sustainability of a Thermal Envelope Renovation Project (TERP) defining economic and environmental impacts according to social aspects of possible TERP's preselected by the reference model.

We want to assist the decision-making process at technic level through the comparison of environmental/economic impact of different DOs. It is why we identify as analysis object assembled systems or building parts analysed in previous research (Appendix A) to an economic/environmental assessment.

These results are social contextualized to determine deviation caused by consumption trends of users and to qualify and assess accuracy and satisfaction level of users on use phase of renovated building and building to renovate.

### 3.1.2 Preview Use

Economic and sustainable assessment must satisfy the principle "the less cost of LCA of BDO's is the most sustainable". This implies reasonable settings between different design options, keeping in mind the three-dimensional balance through promoting sustainable areas more unbalanced.

We propose a method that does not provide us reference values to categorize/qualify interventions but compare them thanks to impacts calculated by quantification model.

## 3.2   Assessment Object Specification

A key aspect to determine the object specification is to establish a Required Life Service (ReqLS) of building based on an Assessment Building Report—according to Spanish Law Royal Decree 7/2015—to ensure structure and foundations aptitude to performer renovation actions. Any building renovation wants to increase service period of building, and it depends on the structure aptitude service report of the Assessment Building Report.

We consider a service period of building of 50 years, according to others research [1–3], establishing a ReqLS of 84 years, finishing this in 2070.

### 3.2.1   Reference Study Period (RSP)

Reference Study Period initiates now the DO's assessment, and it is coincident with the rest of Required Life Service. This let consider environmental and economic impact of design options combinations (DOC) studied.

### 3.2.2   System Limits

Limits to different assembled systems assessed encompass the stages of life cycle defined in Table 3.1. According to environmental and economic dimensions of design option combinations (DOC), we consider product stage, construction and construction waste management (CWM). Complementary, we include on assessment the Module D [...], due the design option study objective goes beyond the limit of considered system—its construction. We structure this information beyond the system in the following:

- $DB_6$ module, use of energy, considering environmental and economic benefits derivatives of TERP through emissions reduction and energy–cost reductions.
- $DB_2$, $DB_3$ and $DB_4$, considering economic impacts of each DO preselected. We do not quantify environmental impact of maintenance operations, due to the difficulty to get consistent information.

**Table 3.1** Sustainability assessment and considered modules

|  |  |  | Environmental assessment | Social assessment | Economic assessment |
|---|---|---|---|---|---|
| Product stage | Raw material supply | A1 | X | NAM | X |
|  | Transport | A2 | X | NAM | X |
|  | Manufacturing | A3 | X | NAM | X |
| Construction process stage | Transport | A4 | X | NAM | X |
|  | Construction–installation process | A5 | X | NAM | X |
| Use stage | Use | B1 | NAM | X | NAM |
|  | Maintenance | B2 | NAM | X | NAM |
|  | Repair | B3 | NAM | X | NAM |
|  | Replacement | B4 | NAM | X | NAM |
|  | Refurbishment | B5 | NAM | X | NAM |
|  | Operational energy use | B6 | NAM | X | NAM |
|  | Operational water use | B7 | NAM | NAM | NAM |
| End-of-life stage | Deconstruction demolition | C1 | NAM | NAM | NAM |
|  | Transport | C2 | X | NAM | X |
|  | Waste processing | C3 | NAM | NAM | X |
|  | Disposal | C4 | NAM | NAM | X |
| Supplementary information beyond the building life cycle | Operational energy use | $D_{B6}$ | X | NAM | X |
|  | Maintenance | $D_{B2}$ | NAM | NAM | X |
|  | Repair | $D_{B3}$ | NAM | NAM | X |
|  | Replacement | $D_{B4}$ | NAM | NAM | X |
|  | Deconstruction demolition | $D_{C1}$ | NAM | NAM | X |

Social dimension assesses usage stage (modules B1–B5), it implies to consider the present situation of building (B1)—before the intervention—and others related to TERP (modules B2–B5)—during the definition of project, execution project and after in execution phase.

## 3.2.3 Building Model. Physical Characteristics

Table 3.2 describes most relevant physical characteristics of building and necessary information to consider face to envelope renovation.

**Table 3.2**  General information of model

| Building | Housing tower |
|---|---|
| Year of construction | 1976 |
| Stories | 12 |
| Functional programme | Four bedrooms, one kitchen, one laundry room, two bathrooms and one living room |
| Occupied floor area (m$^2$) | 5433.80 |
| Occupied floor volume (m$^3$) | 16,301.50 |
| Other electrical gains (W/m$^2$) | 4.40 |
| Occupancy (m$^2$/person) | 0.03 |
| Infiltration (ac/h) | 0.70 |
| Natural ventilation (ac/h) | 4 |
| Windows opening | CTE residential night ventilation |
| Minimum level of illuminance (lux) | 100 |
| Wall construction | – 1/2-foot metric perforated brick |
|  | – Vertical unvented air chamber 4 cm |
|  | – Hollow brick partition simple 4 cm |
|  | – Plaster, low hardness 1.5 cm |
| Glazing construction | – Aluminium window frame (no break) |
| Roof construction | – Waterproofing sheet |
|  | – Formation of slopes and structure |
| Wall $U$-value (W/m$^2$ K) | 1.44 |
| Glazing $U$-value (W/m$^2$ K) | 5.86 |
| Roof $U$-value (W/m$^2$ K) | 1.25 |
| Wall colour | Brick |
| Solar control glazing | No |
| Roof colour | Red |

## 3.3   Scenario's Definition, Indicators Selection and Quantify Methods to Assess Comprehensive Sustainability of TERP

At first, we justify the accuracy of criteria, contrasting them with reference research in this field. Scenario's definition and selection and methods to quantify impacts have been previously by authors [4], and consequently, we present a brief summarized at the second part of this section and a brief summary with remarkable items.

### 3.3.1 Criteria Justification to Develop Scenario and Selection of Indicators from an Environmental Perspective

We select, through quantification model, indicators: (1) abiotic resource depletion potential for elements, expressed by MJ (ADP) and (2) global warming potential, expressed by kg $CO_2$ equiv (GWP).

Complementary, to understand environmental impact of envelope's upgrade, we use indicator TPB $CO_2$ ($CO_2$ time payback), linking emissions savings by measures over envelope with impact of construct them.

The model selected is representative from a city scale view, and it is why considering embodied carbon of constructive solution to renovate thermal envelope is useful to minimize the impact of a unitary intervention in a neighbourhood.

Rivero Camacho et al. [5] underlay the importance of considering carbon footprint of constructive system to implement in energy retrofitting, becoming an essential tool of decision-making.

We consider environmental impacts of constructive solution according to UNE-EN 15978:2012. Following these rules, (1) we justify different scenarios, (2) we define data traceability, and (3) we describe calculations methods.

Different embodied carbon aggregated data are embedded in quantification model, based on building information modelling (BIM) through shared parameters, in line with previous research. Results obtained through this tool are analysed from a LCA perspective.

### 3.3.2 Criteria Justification to Develop Scenario and Selection of Indicators from a Social Perspective

We define a qualification model through the definition of social indicators, being the main categories such as (1) accessibility, (2) adaptability, (3) health, (4) comfort, (5) neighbour's impacts, (6) maintenance and (7) security.

Indoor thermal characteristics of dwellings in use stage are determinant face to contextualize sustainability markers from an environmental and economic perspective. Supporting this idea, Sánchez-García shows how through thermal comfort adaptability of users we can predict saves around 19% and 25% in climate scenarios estimated to years 2050 and 2080, respectively—from a climate change perspective in a A2 hypothesis, according to IPCC. According to this, to consider dynamic set points temperature in future, climate scenarios are an effective tool to estimate consumption reduction in centralized HVAC system. Our functional equivalent does not have this typology of HVAC system, being individual systems. Each user profile determines consumption, and it is the reason why it is strongly needed to know the consumption culture of neighbours to: (1) estimate investment returns and (2) identify indoor renovation opportunities.

### 3.3.3 Criteria Justification to Develop Scenario and Selection of Indicators from an Economic Perspective

We select, through a quantification model, indicators: (1) initial cost (COINIT), (2) maintenance cost ($CO_{MA}$), (3) net present value (NPV), (4) equivalent annual annuity (EAA) and (5) global cost (GC).

Complementary, to understand return investment, we use the indicator economic time payback ($TPB_{\in}$).

### 3.3.4 Environmental Dimension: Selection of Scenarios and Calculation of Environmental Data to Assess the Performance According to UNE-EN 15978:2012

Table 3.3 summarizes selected systems limits, scenarios, indicators, data origin and calculation tool used.

We selected four stages, according to the same system limits that we can summarize in 3:

- Product and construction stage (modules A1–A5).
- End-of-life stage (module C2).
- Supplementary information beyond the building life cycle (module $D_{B6}$).

The first one (modules A1–A5) includes the manufacturing process of the construction systems and materials that are used in the envelope renovation, as well as the construction processes. The information for the stage between modules A1 and A3 is defined by the environmental product declarations (EPD), according to EN 15804. Since not all the evaluated systems have EPD, we use the Technical Approval Documents (DIT, Spanish acronym), which have been used by reference institutions as a starting point to compile their own databases that allow us to know approximately the environmental impacts associated with different construction systems. Similarly, the impacts associated with the construction process are estimated through these databases.

The second one (module C2) is limited to impacts derived from the transport to landfill of CWD generated in the renovation work. We calculate environmental impact through weight and volume of waste generated for each DO, which has been determined on the basis of the waste classification system associated with each construction system according to its EWL code (European Waste List) identified in the reference database [4]. We calculate transport impact depending on the maximum load of the vehicles used (N1 and N2, according to directive 70/156 EEC) and the main fuel of 45% of the transport trucks in the city of Malaga (Diesel), and the $CO_2$ emissions associated with this activity have been calculated according to the type of vehicle (161.2 g $CO_2$/km and 218.47 g $CO_2$/km respectively), according to the statistical tables provided by the directorate general of traffic (DGT) on the vehicle

**Table 3.3** Environmental dimension: system limits, scenarios, indicators, data origin and tools used

| Systems limits | | | Scenarios | Indicators | Data origin | Tools |
|---|---|---|---|---|---|---|
| Product stage | A1–A3 | Raw material supply, manufacturing and transport | Impact asseses of different OD's materials | GWP and ADP | BEDEC[a], DIT and EPD | BIM quantification model |
| Construction process stage | A4–A5 | Transport and construction–installation process | Impact asseses of different OD's construction process | GWP and ADP | BEDEC[a] | BIM quantification model |
| End-of-life stage | C2 | Transport | Energetic cost of transport to landfill of CDW | GWP and ADP | DGT[d] and IDAE | BIM quantification model |
| Supplementary information beyond the building life cycle | D_{B6} | Operational energy use | Energy performance asses after TERP implementation | GWP | Climatic data basis (IPCC) | DesignBuilder thermal model |

[a]BEDEC: product construction database developed by ITEC (from Spanish, Technic Institute of Construction)
[b]DIT: from Spanish, Technical Suitability Document
[c]EPD: Environmental Product Declaration
[d]DGT: the Directorate General of Traffic

fleet [6] and IDAE (from Spanish, Institute for Energy Diversification and Saving) databases for vehicle emissions [6].

Finally, the third one (module $D_{B6}$) calculation procedures for the regeneration proposal are carried out in accordance with the calculation procedure determined by the UNE-EN 15603 standard, relating to the energy efficiency of buildings, overall energy consumption and the definition of energy assessments. The calculation is carried out using the DesignBuilder tool. Climate change repercussions, quantifiable through the period of compensation of $CO_2$ emissions over the useful life of the renovated building, are determined through the calculation of three different time scenarios (2020, 2050 and 2080) which are defined through the morphing of local climate files.

### 3.3.4.1    Calculation Methods and Tools

For DO's environmental impacts calculation of different life cycle module, we use principles established by the UNE-EN 15978 standard. By the determination of indicators and modules mentioned above, we obtain a calculation matrix as follows:

**Equation 3.1**  Matrix expression of the calculation of environmental indicators in different stages of the life cycle of the building renovating actions. *Source* Own elaboration according to UNE-EN 15978

| Quantity of products/processes used in stage $i$ | | | Environmental impact per unit of product/process | | Environmental impact of stage $i$ | |
|---|---|---|---|---|---|---|
| | Database for $a_1$ stage $i$ | Database for $a_2$ stage $i$ | Database for $a_3$ stage $i$ | Database for $a_{...}$ stage $i$ | Database for $a_n$ stage $i$ | |
| $\begin{pmatrix} a_{1,i} \\ a_{2,i} \\ a_{3,i} \\ a_{...,i} \\ a_{n,i} \end{pmatrix} \cdot$ | $\begin{matrix} GWP_{a1,i} \\ ADP_{a1,i} \end{matrix}$ | $\begin{matrix} GWP_{a2,i} \\ ADP_{a2,i} \end{matrix}$ | $\begin{matrix} GWP_{a3,i} \\ ADP_{a3,i} \end{matrix}$ | $\begin{matrix} GWP_{a...,i} \\ ADP_{a...,i} \end{matrix}$ | $\begin{matrix} GWP_{an,i} \\ ADP_{an,i} \end{matrix}$ | $\begin{pmatrix} GWP_{,i} \\ ADP_{,i} \end{pmatrix}$ |

For $i = $ [A1 a A3, A4, A5]

Thus, the quantities of products determined through the measurements of the different DO for the TERP determined according to previous research—EVFB4 + M1, EVH-C3 and EHCP-B2 + M1—(see Appendix B) are multiplied by the impacts associated with them and established per unit area. The joint impact is expressed for stages A1–A5 and C2.

On the other hand, the calculation method for stage $D_{B6}$ is based on the computerized assessment of the energy performance of the building in the different time scenarios.

### 3.3.5 Social Dimension: Selection of Scenarios and Calculation of Social Data to Assess the Performance According to UNE-EN 16309+A1:2015

We define a calculation method with scenarios based on usage stage, contemplating modules from B1 to B7 according to two different situations:

– Building at present
– Building during and after TERP execution.

However, we understand assessment could imply other social dimensions, as the positive impact on the local business and industrial fabric and the job creation, but we will need different data and information modules not included in the reference normative.

We understand the social dimension from users' perspective on usage stage according to mentioned scenarios.

At first, we need to identify indicators related to envelope renovation actions as essential part of a complete regeneration project. Table 3.4 shows preselected modules depending on each scenario:

- **Building at present**

  – We identify indicators to information module B1. We do not consider the rest of modules due to which they do not give us useful information.
  – We propose a list of questions according to the identification of indicators.
  – We assess this indicators through a telematic survey to assess users' interest on renovation building.

- **Building during and after TERP execution**

  – We identify indicators for the following information modules: (1) information module B1—use—to determine TERP accuracy to starting necessities; (2) information module B2—maintenance—to assess impacts of DOC's maintenance; (3) information module B3—repair—to assess impacts of DOC's repair; (4) information module B4—replacement—to assess impacts of DOC's replacement on different TERP phases; (5) information module B5—refurbishment—to asses impacts of all works related to the complete renovation project,

**Table 3.4** Social scenarios and information modules selected

| Scenarios | Information modules to assessment | | | | | | |
|---|---|---|---|---|---|---|---|
| | B1 | B2 | B3 | B4 | B5 | B6 | B7 |
| Building at present | X | NAM | NAM | NAM | NAM | NAM | NAM |
| Building during and after TERP execution | X | X | X | X | X | X | NAM |

NAM: not assessed module

and information module B6—operational energy use—to determine whether proposed DOC satisfies energy savings expected.

This proposal method wants to contextualize results according to these two phases—before and after TERP—and adapt results according to users' satisfaction grade, accuracy of solutions adopted and execution impacts on neighbours. Nevertheless, we focus on the present, before TERP, in order to establish influence of social dimensions on results.

We consider indicators considered on UNE-EN 16309+A1:2015, selected those related to assessment objective and others partially related to Thermal Envelope Renovation Project. Appendix B lists this selection and the impact allocation.

### 3.3.5.1  Indicator's Identifications, Questions and Model Survey

The UNE-EN 16309+A1 standard sets out a lists of indicators and a methodology for the evaluation of the B1 information module that support the identification of the indicators.

Tables on Appendix B, among all the indicators that can be considered, are related to the object of study. Column "a" identifies the relevance or not of the indicator—the table is a selection of relevant indicators, which is why column "a" is relevant. Columns "b" and "c", on the other hand, determine whether or not there is a national, regional or European requirement applicable to the indicator in question and identify this requirement. Finally, columns "d", "e" and "f" show whether or not, at the time of the assessment, measures have been taken to exceed the requirements identified in column "c" and, if not, provide a value for the indicator or specify the measures to be taken to meet the requirements and provide the regulatory framework that relates to aspects reflected in columns "c" and "e". In summary, and in relation to the reference standard, each of the above-mentioned columns expressly answers the following questions and/or specifications:

(a)  Is the aspect relevant to the design of the object of assessment? (Yes/Not Relevant/NAI).
(b)  Is there an applicable national, regional or European requirement?
(c)  Specify the minimum requirement according to the preferred regulation or, if not regulated, provide details of any requirements in the client's specifications (if none, state "No").
(d)  Have measures been taken to overcome the requirements presented in c? (Yes/No).
(e)  Provide a value for the indicator or briefly specify the measures taken to achieve the stated performance.
(f)  Provide references to relevant documentation to support what is stated in column "e".

It should be noted that the assessment of information module B1 of the planned and implemented status of the regeneration project shall satisfy the apparent shortcomings resulting from the assessment of the same information module for the current status of the project. By this, it is meant that the following tables do not contain information on the assessment of indicators at the use stage for the planned/implementation state of the PRET, as it is understood that the only difference with respect to the same assessment for the current state of the building is the verification in column "d" that measures have been taken to overcome or achieve the requirements presented in column "c", which have already been identified.

Appendix B tables list the **38 selected indicators**.

Attending to the different categories of social behaviour of the standard proposes, all the proposals have been considered, and according to the criteria summarized below, the relevant indicators have been considered or not with a view to the social evaluation of the potential strategies for improving the thermal envelope of building to be proposed, as part of a building regeneration project focused on the renovation of the envelope.

- **Accessibility**

Accessibility of buildings is an issue that will inevitably have to be addressed in any building renovation projects that must be considered. With regard to this issue, the reference standard provides very useful indicators that will have to be addressed and verified in order to propose reasonable adjustments to buildings to improve their accessibility. Of all the indicators proposed by the standard, the most relevant is the ubication, dimensions and ease of operation of lifts, especially because this will be a service that must be guaranteed in the course of any regeneration work that may be carried out. It has also been verified that the dimensions of the lift do not guarantee its accessibility, which could be the subject of an analysis and proposal of alternatives in the framework of a complete regeneration project for the building.

- **Adaptability**

From a social assessment perspective, the analysis of this category is crucial, since the envelope regeneration actions, understood within the framework of a building regeneration project, should be able to satisfy the changing needs of use of the users and/or owners of the dwellings. This is one of the most important aspects in the survey of the building's residents, as it is crucial to know their interest in functionally adapting their homes by means of interior redistribution and/or improvement of installations (see Appendix D). At the level of identification of these indicators, it is observed that none of them are satisfied and that the interior conditions of the dwelling do not adapt to the minimum requirements at a regulatory level, which, as can be observed in phase 4, does not translate into an express need on the part of the residents to modify the interior conditions of their dwelling.

This category acquires a decisive relevance at the end of the research in view of the change in users' needs in relation to the functionality of their dwellings as a result of the current health crisis generated by COVID-19 and which has already been referred to.

- **Health and comfort**

Thermal, indoor air quality, acoustic and visual comfort characteristics are particularly related to renovation actions on thermal envelope. Parallelly, this type of interventions can be an opportunity for visual comfort improvement, by means of new openings on the façade that also improve the natural lighting of indoors.

– Maintenance

Considering the influence of maintenance and maintainability operations on the daily life of the users of the dwellings is relevant when making decisions on the type of solutions to be implemented in order to improve the health and comfort inside the dwellings. Thus, the adoption of regeneration measures that involve costly maintenance and/or that could have a negative impact on the activity of the users—for example, because they would have to be carried out from inside the dwellings—should be discarded in favour of low-cost renovation actions that can be carried out without any incompatibility of use and/or visual impact on dwellings.

– Safety and security

Some of the indicators included in this category are not considered, given climatic conditions of the buildings location. However, behaviour in earthquakes and fire events could be included by means of consideration of normative regulations such as Basic Document of Structure Security and Basic Document of Security in case of Fire, both of them included in Spanish Building Technical Code.

– Impacts on the neighbourhood

This social behaviour subcategory is particularly relevant in execution phase. We will not evacuee the building during the works of renovation, which inserts a complexity in the construction process that will have to be considered. The subcategories of social behaviour collected (noise, emissions, glare/shading and shocks/vibrations) allow us to identify relevant aspects to be considered to reduce their impact on the neighbourhood, taking the relevant measures to reduce their effects.

The identification of 32 social impact indicators related with TERP allows:

– To know deeply needs of neighbours for this renovation actions (module B1 of building at present) and
– To determine, through the assignment of influences established—see Appendix B—those impacts to be considered throughout the life cycle of the building and the measures to avoid or minimize them, as well as the value proposals for each of them.

From the analysis of the measures to avoid social impacts throughout the life cycle with regard to the TERP execution, it can be concluded that the maintenance organization plan for systems must be implemented which allows reduction of impacts. Likewise, it is found that, as far as possible, it is adequate to implement simultaneously TERP and functional indoor transformations, which allows reducing impacts on the neighbourhood throughout the life cycle.

All this highlights the interest of the qualitative model proposed, which is defined through the selected indicators, and which is evaluated before the execution of the actions, for the definition of the execution project and for the maintenance of building.

Finally, indicators that allow supporting the decision-making of TERP are highlighted for their relevance as follows:

- **Indicator 7.2.2**. The location, dimensions, and ease of operation of lifts—associated with the intervention and in module B5. As far as possible, actions on the building envelope should be carried out simultaneously with works inside the dwellings, given that the provision of auxiliary means for the renovation of the envelopes (scaffolding and/or lifting systems for external materials) means that they do not depend exclusively on the communication elements of the tower of dwellings, thereby also improving safety during the works.
- **Indicator 7.3**. Adaptability—associated with the intervention and in module B5. Redistributing internal spaces while transforming the vertical envelope makes it possible to integrate new openings in the façades that improve the interior/exterior relation, as well as the orientation and lighting of the rooms.
- **Indicator 7.4.6**. Outdoor space—associated with the intervention and in module B5. The recovery of outdoor spaces that have been lost and the extension of existing ones are design strategies that can be studied and incorporated into the PRET.
- **Indicator 7.4.5.1**. Visual comfort features—associated with the intervention and in module B5. Transforming the existing openings and defining them as perfectible elements that can be intervened on throughout the remaining useful life of the building is an important element for the definition in the design of a system of modules and rhythms of the façade that can respond to the different indoor redistributions.

Considering the selected 32 indicators, we make a list of question to carry out a survey. Users' perception about renovation necessity of building has been estimated as responsible of 44% of building renovation operations [7]. To carry out the survey, the indicators have been synthesized into very specific questions that have been divided into different blocks (Table 3.5).

The survey must be carried out at the level of the building where the intervention takes place. However, in order to collect a broader representative sample in relation to the neighbourhood, for the application of the qualitative model, the neighbours of different housing towers that respond to the functional equivalent are considered for the application of the qualitative model.

**Table 3.5** Survey results in the building for the selected indicators in relation to the culture of use of neighbours and issues related to the potential TERP to be implemented in the building

**Starting data**

| | | North | South | East | West |
|---|---|---|---|---|---|
| Orientation (living–dining room) | 24% | 32% | 21% | 24% | |
| Flat | | 12% | 01 / 12% | 02–11 / 76% | 12 / 12% |
| Owner or lessee | | | | | 100% |
| Age profile of home users (years) | | | 20–45 / 12% | 41–65 / 41% | >65 / 47% |

*Adaptability*

| | | | | |
|---|---|---|---|---|
| 2/1.1 | How many people do live in the house? | | | 2.4 |
| | Evaluate from 0 to 5 and five being the highest score. How satisfied you are with the characteristics of each of the rooms in the house? | | Hall | 3.9 |
| | | | Kitchen | 4.3 |
| | | | Laundry | 4.2 |
| | | | Dining room | 4.4 |
| | | | Toilet | 3.9 |
| | | | Bedrooms | 4.3 |
| | | | Bathroom | 4.1 |
| | | | Terrace | 4.2 |
| | | | Mean | 4.2 |
| 2/1.2 | Does the house fit your current needs? | 100% | Y | 0% | N |
| | Does the house adapt to the changes that occur in your life? | 94.1% | Y | 5.9% | N |
| | Have you thought about changing your home? | 11.8% | Y | 88% | N |
| | Have you thought about modifying the house where do you live? | 41% | Y | 59% | N |
| 2/1.3 | Does the house accommodate to new technology? | 97% | Y | 3% | N |
| 2/1.4 | Have you thought of a different use for this dwelling compatible with the residence? | 0% | Y | 100% | N |

(continued)

**Table 3.5** (continued)

Starting data

*Health and comfort*

| Code | Question | Option A | Option B | Option C | | Y | | N |
|---|---|---|---|---|---|---|---|---|
| 3/1.1 | Do you get cold inside the house in winter? | | | | 24% | Y | 77% | N |
| | What measures are taken to avoid it? | Heater 58% | HVAC 8.3% | Cloth 33% | | | | |
| | Do you pass heat inside the house in summer? | | | | 55.9% | Y | 44.1% | N |
| | What measures are taken to avoid it? | Fan 55.0% | HVAC 40.0% | Cloth 5.0% | | | | |
| 3/1.2 | Do you feel excessive humidity in your home? | | | | 8.8% | Y | 91.2% | N |
| 3/1.3 | Do you ventilate the house daily? | | | | 97.1% | Y | 2.9% | N |
| 3/1.4 | Do you feel comfortable in the activities you do inside? | | | | 97.1% | Y | 2.9% | N |
| 3/1.5 | Do you consider it annoying to change your clothing? | | | | 5.9% | Y | 94.1% | N |
| 3/2.2–3/4.2 | Can you control the air conditions in the rooms? | | | | 88.2% | Y | 11.8% | N |
| | If yes, what systems do you use? | 100% | Manually | | | | 0% | Auto |
| 3/2.3 | Do you have any system to know the internal air conditions? | | | | 14.7% | Y | 85.3% | N |
| 3/5.5 | Do you consider adequate the acoustic insulation? | | | | 52.9% | Y | 47.1% | N |
| 3/5.6 | Does the house have enough natural light? | | | | 100% | Y | 0% | N |
| 3/7.2 | Does the house have awnings? Do you use them? | 5.9 Never | 35.3% Daily | By season | 8.8% | Y | 50.0% | N |
| 3/8.2 | Do you consider that the interior height of your home is adequate? | | | | 97.1% | Y | 2.9% | N |
| 3/8.6 | Do you have a terrace with enough surface for the use you make of it? | | | | 94.1% | Y | 5.9% | N |

(continued)

**Table 3.5** (continued)

**Starting data**

| | Question | | | | | | | |
|---|---|---|---|---|---|---|---|---|
| 3/11.1 | Do you consider that there is a lot of external contamination and that affects the cleaning needs of the home? | | 58.8% | Y | 41.2% | N | | |
| 3/11.2 | Perceiving bad odours in your home derived from the building's facilities? What facilities specifically? | 26.5% Y | 2.9% Y | 8.8% | 61.8% | N | | |
| | | Sanitation | Shunt vent. | Both | | | | |
| 3/11.3 | Do you have air conditioning in the house? | | 82.4% | Y | 17.6% | N | | |
| | If yes, do you use it very often? | 0% Always | 18% | Summer days | 82% | Hot days | | |
| 3/12.1 | Do you consider that there is excessive light pollution in your environment? Is it difficult to darken the house at night? | 11.8% | Y | 8.8% | Y | 79.4% | N | |
| | | Difficult darkening | | No difficult darkening | | | | |
| 3/12.2 and 3/12.3 | Do the surrounding buildings shade the home? Would you claim that you don't have enough direct light? | 2.9% | Y | 11.8% | Y | 85.3% | N | |
| | | More light | | No more light | | | | |
| 3/14.1 | Would you say that proper house maintenance is carried out? How often are repairs made? | 2.9% Y | 55.9% | 5.9% | N | 35.3% | | |
| | | | Monthly | Annually | | Monthly | Annually | |

**Security**

| | Question | | | | | | | |
|---|---|---|---|---|---|---|---|---|
| 4/15.1 | In your experience, would you say your house adequately resist the torrential rains of recent years? Do you notice any damage/humidity/leaks? | 5.9% Y | 8.8% | Y | 2.9% | 82.4% | Y | 0.0% N |
| | | Failures in facade | Failures in facade | Failures in hollows | No failures | | | |
| 4/15.2 | Have you had problems with rainwater leaks? | | 2.9% | Y | 97.1% | N | | |
| 4/18.3 | Have you had a problem with humidity/condensation/leaks on the facade? | 5.9% Y | 0.0% | Y | 0.0% | 94% | N | |
| | | Humidity | Condensations | Leaks | | | | |
| 4/19.1 | What solar control measures do you use most frequently? | 71% | 15% | 15% | | | | |

(continued)

**Table 3.5** (continued)

Starting data

|  |  | Blinds |  | Curtains |  | Awnings |  |
|---|---|---|---|---|---|---|---|
| 4/19.2 | Have you exchanged windows for better features in the last 10 years? | 68% |  | 53% | Y | 47% | N |
| 4/19.6 | Do you consider the orientation of the house good? Do you save energy for it? |  | Y | 2.9% | Y, no | 29% | N |
|  |  |  |  | Savings |  |  |  |

### 3.3.6 Economic Dimension: Selection of Scenarios and Calculation of Economic Data to Assess the Performance According to UNE-EN 16627 and UNE-EN 15459-1

According to principles of standard to life cycle cost (LCC) calculations, we determine selected scenarios, indicators, data origin and tools. Table 3.6 summarizes contents of this section. Module information data origin is different, so that in this condition we divide them.

We selected four stages, according to the same system limits that we can summarize in 3:

- Product and construction stage (modules A0–A5).
- End-of-life stage (modules C2, C3 and C4).
- Supplementary information beyond the building life cycle (modules $D_{B2}$, $D_{B3}$, $D_{B4}$, $D_{B6}$, $D_{C1}$).

The first one (modules A0–A5) includes phases prior to the A1 information module, because it is necessary to know what economic resources are available before the TERP development. This is why we must have a forecast of incentives and costs linked to the renovation actions. The European next generation funds, through the specific national programmes, allow different types of aid to be available at the State and Autonomous Community levels. Table 3.7 shows the main aspects to be considered from an economic perspective in the preconstruction phase (information module A0). Likewise, aspects such as the costs of licences for the actions proposed by TERP, professional fees and increases in the base bidding budget (industrial profit and indirect costs) have been considered for the preparation of this table.

We have chosen to work with the hypothesis where renovating actions are covered by the programme for the promotion of energy efficiency and sustainability improvement in housing promoted by the Spanish Ministry of Development. The requirements and limitations established in this programme respond to the actions' characteristics, users' profile and building's typology, also entailing the most favourable situation with regard to the financing of the actions.

In order to calculate the amounts that could be accessed, it is necessary to previously determine the value of the Public Indicator of Multiple Effect Income (IPREM), a common indicator of the different subsidy and aid programmes for the regeneration of buildings in order to establish maximum amounts, considering those family incomes that do not exceed this indicator by a certain number of times. The results for this indicator in 2019 established a monthly IPREM of 537.84 €; an annual IPREM (12 payments) of 6454.03 € and an annual IPREM (14 payments) of 7519.59 €. In the case of the two programmes for which the building under study could be eligible, the family income would have to be below the thresholds identified in Table 3.8 in order for these subsidies to be increased.

Data have been calculated through the surveys, where respondents were asked to identify the range of average gross monthly household income. The monthly indicator

**Table 3.6** System limits, scenarios, indicators, data origin and tools to the economic assessment

| System limits | Scenario | | Scenario | Indicator | Data origin | Tools |
|---|---|---|---|---|---|---|
| Product stage | A0 | Preconstruction | Previous economic quantifications to TERP (cost, incentives, aids, etc.) | € | Administration | BIM[a] + PRESTO[c] |
| | A1–A3 | Raw material supply, transport and manufacturing | Economic quantification assessment of system/material to renovate thermal envelope and definition of a material cost budget | € | CYPE[b] | BIM[a] + PRESTO[c] |
| Construction process stage | A4–A5 | Transport and construction–installation process | Economic quantification assessment of human resources to DOC construction and definition of a manpower budget | € | CYPE[b] | BIM[a] + PRESTO[c] |
| End-of-life stage | C2 | Transport | Economic quantification assessment of CDW transport to a required landfill | € | CYPE[b] | BIM[a] + PRESTO[c] |
| | C3 | Waste processing | Economic quantification of CDW treatments and disposal | € | Administration | BIM[a] + PRESTO[c] |
| | C4 | Disposal | | | | |
| Supplementary information beyond the building life cycle | D[B6] | Operational energy use | Economic quantification of building consumption in operational phase | € | REE[e] and OMIE[d] | BIM[a] + PRESTO[c] |
| | D[B2] | Maintenance | Cost maintenance, repair, and replacement estimations in normal usage conditions | € | CYPE[b] | BIM[a] + PRESTO[c] |
| | D[B3] | Repair | | | | |
| | D[B4] | Replacement | | | | |

(continued)

**Table 3.6** (continued)

| System limits | Scenario | | Scenario | Indicator | Data origin | Tools |
|---|---|---|---|---|---|---|
| | $D_{C1}$ | Deconstruction demolition | Cost deconstruction and demolition estimations of DOCs | | | |

[a]BIM: building information modelling
[b]CYPE: architecture, engineering and constructive software
[c]PRESTO: measurements and budgets software
[d]OMIE: Iberian electric market operator
[e]REE: Spanish electric net

**Table 3.7** Quantification and cost/income applicability associated with TERP to the study case

| Cost/income | Data origin and/or concepts | Apply? | % | Min–max amount |
|---|---|---|---|---|
| Incentives or subsidies available to promoters/users of renovation measures | Programme for the promotion of energy efficiency and sustainability improvement in housing promoted by the Spanish Ministry of Development | Y | 40% | 384,000 € 672,000 € |
| Costs of licences to be applied for | Tax ordinance. Fees for urban development actions. Amount depending on the type of action | Y | – | – |
| Professional fees | Professional associations | Y | | – |
| Basic offer budget increments | General cost | Y | 13% | – |
| | Industrial profit | Y | 6% | – |
| | Value-added tax | Y | 21% | – |

**Table 3.8** Aids increment benchmarks

| Inventive | Conditions | Benchmarks | | |
|---|---|---|---|---|
| Programme to promote the improvement of energy efficiency and sustainability in housing | A maximum of €8000 per dwelling is established, which can be increased by 75% | 3.0 × IPREM | Monthly | 1613.52 € |
| | | | Annually (12p) | 19,362.09 € |
| | | | Annually (14p) | 22,558.77 € |
| Andalusia Housing and Rehabilitation Plan 2016–2020 | Set at up to €7200 if the family income is less than 2.50 times the IPREM | 2.50 × IPREM | Monthly | 1344.60 € |
| | | | Annually (12p) | 16,135.08 € |
| | | | Annually (14p) | 18,798.98 € |
| | Set at up to €8800 if the family income is less than 1.50 times the IPREM | 1.50 × IPREM | Monthly | 806.76 € |
| | | | Annually (12p) | 9681.05 € |
| | | | Annually (14p) | 11,279.39 € |

was considered to be the most appropriate for inclusion in the questionnaire due to the simplicity of the question's approach. Thus, the thresholds of €800, €1300 and €1600 were identified in the questionnaire. The results show that the income of 47.10% of the residents surveyed is below the first threshold, 17.60% have a monthly income of less than €1300 per month, compared to 8.80%, whose income is between €1300 and €1600. The household income of 26.50% of the remaining residents is above €1600. With all of the above, in order to draw up a profile of users requesting aid for the rehabilitation of buildings, we can determine that, in relation to the results of the surveys carried out, 73.50% of the residents surveyed are in a position to receive aid that would be increased to €8000 per dwelling. These results are extrapolated to the building under evaluation, which determines that around 34 residents could benefit from an increase in aid, which could cover up to 75% of the corresponding investment, which would have to be managed by means of a separate dossier.

As the consideration of the issues referred to for the preconstruction stage (A0) is decisive from the scope intervention perspective and the approach of different phases of action, they will have to be prefixed, agreed and must structure the decision-making in the rest of the assessments.

However, public–private formulas should be tested to cover the total amount of work to be carried out. This leads us to consider promotional figures that can range from the organization of neighbourhood cooperatives for the regeneration of buildings, to the participation of entities in charge of promoting and coordinating these actions, which may or may not have the support of financial entities that have an interest in the regeneration of the buildings or the urban environment in which they are inserted. In any of the formulas that are determined to be the most suitable, and even in the coexistence of several, it will have to be established with sufficient guarantees that the interests of the agents involved are far removed from real estate speculation. This is only possible if, through the concession of lines of financing for the actions, those involved in the management of these interventions are required to ensure that there is an ethical purpose in them, and that they do not provoke phenomena related to the gentrification of the neighbourhood and/or the deterioration of the existing social fabric.

Costs associated with the construction stages ranging from modules A1 to A5 have to be determined. As shown in Appendix C, the main source for obtaining this data is the existing databases at the regional level. The application of these costs associated with the measurements generated by the Revit model yields the results analysed below, which also include the non-annual costs and revenues obtained in the preconstruction stage. As can be seen, the economic incentives available to the community of owners to meet the costs of the works cover 40% of the total material execution budget.

The second one (modules C2, C3 and C4) includes the cost of CDW transporting generated at renovation sites to locations for processing and/or storage (C2). Similarly, costs for the treatment of CDW, subject to a number of municipal charges (C3 and C4), are also covered.

The third one stages beyond the system boundary (modules $D_{B2}$, $D_{B3}$, $D_{B4}$, $D_{B6}$, $D_{C1}$) included scenarios during the useful life of renovated building.

Due to the complexity, modules $D_{B2}$, $D_{B3}$ and $D_{B4}$ have been estimated through a decennial maintenance module developed by Spanish software for architecture, engineering and construction (CYPE). In this way, we will be able to determine the estimated ten-year maintenance cost of each of the proposed regeneration operations, as well as the cost of all of them as a whole.

Module $D_{B6}$ has been considered by means of economic savings through the estimation of the consumption calculated by DesignBuilder tool throughout the life cycle of the building for the different time scenarios based on morphed local climate files, in order to determine the impact of climate change on the payback period of the economic investment.

Unpredictability of energy prices due to different relevant recent events (COVID-19 and Russian-Ukrainian war) causes deviation of results difficult to consider, it is why we focus on 2019 prices, which were stables. In order to determine the economic costs, we will use the Report on Regulated Energy Prices prepared annually by the Institute for Economic Diversification and Saving, under the Ministry for Ecological Transition, based on the calculation methodology expressed in the Royal Legislative Decree 216/2014 (Spain, 2014c). The real price values are difficult to determine, as the energy services commercialization companies do not provide sufficiently clear methods for calculating the tariffs and concepts that are difficult to understand and applied in the invoices. By using this data source and the aforementioned methodology, we have a data source whose implementation is on the rise and which, at present, is applicable to those households that have a smart meter, which is why this method of determining costs is considered appropriate, given that in the reference building these reading devices have already been installed.

The hypothesis for calculating these costs is based on the premise of having low-voltage access tolls that do not set timetable discrimination, which means that we are in a 2.0A type of tariff, thus determining the monthly costs shown in Table 3.9. However, these costs fluctuate daily, and their calculation is also legislated through RD 216/2014, but different entities express these data through their pages (OMIE, 2019; REE, 2019b), given the complexity of the calculation, which is subject to multiple economic variables.

The determination of the costs of eliminating ($D_{C1}$) the DOCs is purely indicative and is used to compare different design options and not so much to inform those who face the investment of the costs arising from these actions outside the life cycle (homeowners). It should be noted that, in the hypothesis of a single ownership of the building, this indicator would be relevant for the determination of the costs of demolition and disposal of the building.

In this way, considering different tools available to users for the determination of the costs associated with energy consumption, the different PVPC values over the last

**Table 3.9** Voluntary prices for small consumers (PVPC), according to Iberian electric market operator (OMIE) and Spanish electric net (REE) information

| 2.0A tariff | Access toll | Commercialization cost | | Energy cost |
|---|---|---|---|---|
| | Power term €/kW and month | Energy term €/kWh and month | Power term €/kW and month | Energy term €/kWh and month |
| | 3.1702855 | 0.044027 | 0.25942 | 0.001647 (2014) |
| | | | | 0.001970 (2015) |
| | | | | 0.001589 (2016) |
| | | | | 0.000557 |

five years before both destabilize events. As of the date of the current investigation, a voluntary price for the reference small consumer is determined at 0.113466 €/kWh.

### 3.3.6.1   Calculation Methods and Tools

We develop indicators according to reference standards. DOC's aggregate cost calculation throughout the lifespan of building after the interventions constitutes the basis for determining cash flow of the investment to be made (CF).

Thanks to net present value (NPV), we can determine the current value of the assessed TERP throughout the RSP, enabling to compare it with other alternative projects. Standard indicates that we must consider a real discount rate of 3% to compare different projects.

$$\text{NPV} = -\text{CO}_{\text{INIT}} + \sum_{t=1}^{n} \frac{\text{CF}_t}{(1 + \text{RAT}_{\text{disc}})^t} \tag{3.2}$$

where

$\text{CO}_{\text{INIT}}$   the initial investment to cover the costs of the works, including incentives and aid
$\text{CF}_t$   the income and costs during the study period associated with the renovation project
$t$   number of periods considered
$\text{RAT}_{\text{disc}}$   discount rate.

Costs and income throughout the RSP are determined comparing the costs of the building in its current state—under the current premises of maintenance—with costs associated with maintaining the TERP and the incomes derived from energy savings. The level of aggregation of costs and income is a function of the level of detail available, the sources of which were specified at the beginning of this section.

$$\text{CO}_{\text{INIT}} = C_{\text{pem}} + C_{\text{ht}} + C_{\text{tm}} - S \tag{3.3}$$

where

$C_{\text{pem}}$   material execution budget costs (without VAT)
$C_{\text{ht}}$   technical fee costs (without VAT)
$C_{\text{tm}}$   municipal fee costs
$S$   subsidy granted by the administration to cover part of the construction costs.

Determine the equivalent annual annuity (EAA) as from DOCs imply to know the equivalent income that would be produced throughout the RSP, we use the following expression:

$$EAA = NPV \times \left[ \frac{RAT_{disc}}{1 - (1 + RAT_{disc})^{-t}} \right] \qquad (3.4)$$

where

NPV      net present value
$t$        number of periods considered
$RAT_{disc}$  discount rate.

General cost considers TERP execution different costs. Within the annual costs, we consider the energy and maintenance costs of DOCs. Updated prices of the components and/or services will be established for each of the costs listed as well as the costs for the emission rights according to the data collected by the online platform carbon pricing dashboard [8], which establishes a cost of $16.56 per ton of $CO_2$ emitted. Likewise, removal and dismantling costs are considered at the end of the building expected useful life.

$$GC = CO_{INIT} + \sum_i \left[ \sum_{i=1}^{TC} \begin{array}{c} CO_{a(i)}(j) * (1 + RAT_{xx(i)}) \\ +CO_{CO_2(i)}(j) * D\_f_{(i)} + CO_{f\,disp(TLS)}(j) - VAL_{f in(t_{TC})}(j) \end{array} \right] [\text{€}] \qquad (3.5)$$

where

$CO_{INIT}$           initial investment cost
$CO_{a(i)}(j)$         annual cost for year $i$ of component or service $j$
$(1 + RAT_{xx(i)})$  evolution of prices for year $i$ of the component or service
$CO_{CO_2(i)}(j)$     cost of $CO_2$ emissions for measure $j$ during year $i$
$D\_f_{(i)}$            discount factor of year $i$
$CO_{f\,disp(TLS)}(j)$  final disposal cost for decommissioning
$VAL_{f in(t_{TC})}(j)$  residual value of the assembled system $j$ for the year TC
TC                calculation period.

The investment return is the number of necessary years to compensate the initial investment with energy–cost reduction. We consider the initial reference cost for the building in its current state as 0, since no cost associated with the non-adoption of any measure is expected, it implies that the following expression is the same as the value nominal current, or what is the same, and investment returns will occur when the NPV is equal to zero.

$$\sum_{t=1}^{TPB} CF_t * \left( \frac{1}{1 + RAT_{disc}} \right)^t - CO_{INIT} + CO_{INIT\,ref} = 0 \qquad (3.6)$$

where

TPB     the last year of the payback period (when the formula equals zero)
$t$         number of years from the starting year

$RAT_{disc}$     the discount factor
$CO_{INIT}$     initial investment costs
$CO_{INIT\,ref}$  initial investment costs for the reference case (0 for the option of not renovate the envelope)
$CF_t$          difference in annual costs (difference in cash flow) between the renewed building and the building without any action on it.

## 3.4   Methods Conclusions

The methodology defines and proposes system boundaries and scenarios for assessment in coherence with the functional equivalent, selecting those stages of the life cycle of the retrofitted building for which information is currently available. At constructive level for TERP, the feasibility of integrating into a quantification model is verified.

- Environmental impacts, defined through modules A1–A5 (product and construction process stage), module C2 (transport of CDW, belonging to the end-of-life stage) and module $D_{B6}$ (operating energy use), through which the benefits beyond the system boundary derived from energy savings throughout the LCA that are achieved through the TERP are determined.
- The economic impacts, defined by modules A0–A5 (costs associated with the production and construction of the solutions), C2–C4 (costs associated with the transport, treatment for reuse, recovery and recycling and disposal of the CDW associated with TERP execution) and the benefits after the implementation of the measures originate from the economic savings throughout the LCA (modules $D_{B6}$), as well as the costs associated with the maintenance, replacement and/or repair of these measures (modules $D_{B2-B4}$).

Regarding the developed scenarios for the qualitative model, we define a users' profile before any intervention is planned, in order to know their interests on building's transformations to contribute by means of improvement to aspects such as adaptability, health and comfort and safety. For this reason, two different scenarios are considered:

- Building at present, where module B1 is evaluated in order to ascertain the interests of the users in relation to the transformations to be carried out on the building with regard to the renovation of the building envelope.
- Building during and after TERP execution, where the adequacy of the strategy is verified through the evaluation of module B1 and the remaining use modules (B2–B6, except module B7 for in-service water use) are evaluated. The determination of these modules is established through the allocation of influences on each other, which establishes a different consideration of each of the indicators for each of these modules.

The indicators for economic and environmental dimensions described in this phase allow the comparison of different DOCs. The quantitative model allows the following impact indicators to be obtained in relation to these dimensions for the TERP according to the DOC being assessed:

- $CO_{INIT}$ (€). Initial cost of the proposal.
- $CO_{MA}$ (€/year). Cost of maintaining the DOCs over the remaining lifetime. These costs are updated throughout the LCA in relation to the established financial parameters.
- NPV (€). Net present value of the proposal determines the present value of the proposal in relation to the income and costs that occur throughout the RSP, verifying the contribution of value as long as the value is positive.
- EAA (€). Equivalent annual annuity establishes the uniform annual return on the investment in the transformation of the building.
- GC (€). Overall cost shows updated information on the total economic costs associated with the DOC considering all the stages seen, including the decommissioning of the assembled systems.
- $TPB_€$ (years). Payback period of the investment, allowing to know the year from which the intervention generates an economic value through the energy savings that the implementation of the actions allows.
- GWP (kg $CO_2$ equiv). Global warming potential—also identified as embodied carbon—quantifies the environmental impact associated with the implementation of the measures on the envelope.
- TPB $CO_2$ (years). Emission offset period relates the environmental impact of the DOC to the savings in emissions derived from the implementation of the measures, making it possible to know from which year onwards real savings in emissions are produced.

On the other hand, the identification of indicators for the social assessment of the current state (module B1) leads to the selection of 32 indicators that are related to the actions that are proposed at the envelope level. Through the assignment of influences of some modules on others throughout the stage of use of the building, these indicators are contextualized for each of the modules (see Appendix B).

The evaluation of indicators belonging to module B1 for the current state of the building makes it possible to foresee the degree of intervention on the building, while identifying current consumption trends in relation to the health and comfort category, being in this respect decisive for the estimation of emissions and consumption savings throughout the LCA:

- The consumption trend of users in relation to the air conditioning of their homes.
- The satisfaction degree of residents with their home and the interventions they have carried out on the thermal envelope of the building.

With regard to the second of the above aspects, it should be noted that, as one of the collateral effects of the global pandemic generated by COVID-19, a significant number of residents have detected during the months of confinement the shortcomings of the dwelling in relation to a more prolonged use of the same, which is currently motivating the carrying out of adaptation and adaptation works in the different spaces of the dwelling that had not been foreseen to date.

The indicators assessment relating to modules B2–B6 for the planned and current state of the TERP is carried out by means of verification sheets that have been described in this phase and that constitute a guide for decision-making that the designer will have to integrate into the TERP, being also a document that has to be added to the book of the regenerated building in order to monitor the indicators.

# References

1. Oregi X, Hernandez P, Hernandez R (2017) Analysis of life-cycle boundaries for environmental and economic assessment of building energy refurbishment projects. Energy Build 136:12–25. https://doi.org/10.1016/J.ENBUILD.2016.11.057
2. Ortiz O, Bonnet C, Bruno JC, Castells F (2009) Sustainability based on LCM of residential dwellings: a case study in Catalonia, Spain. Build Environ 44(3):584–594. https://doi.org/10.1016/j.buildenv.2008.05.004
3. Mercader-Moyano P, Ramos-Martín M (2020) Comprehensive sustainability assessment of regenerative actions on the thermal envelope of obsolete buildings under climate change perspective. Sustainability 12(14):5495
4. CYPE (2018) Construction price generator, Spain. Available online: http://www.generadordeprecios.info/#gsc.tab=0. Accessed 1 May 2022
5. Rivero Camacho C, Pereira J, Gomes MG, Marrero M (2018) Huella de carbono como instrumento de decisiónen la rehabilitación energética. Películas de control solar frente a la sustitución de ventanas. Rev Hábitat Sustent 8:20–31
6. Marteinsson B (2005) Service life estimation in the design of buildings: a development of the factor method. PhD, Kungliga Tekniska Hogskolan of Sweden
7. IDAE (2017) Guide to passenger cars for sale in Spain with fuel consumption and $CO_2$ emissions. Available online: http://coches.idae.es/guia-emisiones-consumos. Accessed 1 May 2022
8. Spanish Directorate-General for Traffic (2018) Vehicle fleet-yearbook 2018. Available online: http://www.dgt.es/es/seguridad-vial/estadisticas-e-indicadores/parquevehiculos/tablas-estadisticas/2018/. Accessed 1 May 2022

# Chapter 4
# Method Application

**Abstract** The application of the method to a building in the Mediterranean city of Malaga allows a series of results to be obtained which are included in this section with the aim of obtaining a series of conclusions which are set out at the end of the section.

**Keywords** Results · Case study · BIM · Sustainable construction

## 4.1 Calculations Through Quantitative and Qualitative Models

This section shows the results of the application of the methodology for the TERP of a representative building of a larger urban reality. To this end, the results analysed are the result of differentiated models:

- **Quantitative model.** The environmental and economic assessment dimensions are based on this model and responds to the physical characteristics of a standardized proposal for the improvement of the elements affected by the intervention. This quantitative model is developed in two phases:

  - Initial quantitative model (IQM). It allows determining the impacts of the environmental and economic indicators of different sets of DOCs over the remaining lifetime of the building and relates them to the theoretical savings for that period of time.
  - Final quantitative model (FQM). Part of the IQM, integrating the results of the qualitative model in terms of user consumption trends and the degree of user intervention on the envelope, which allows the calculation of deviations in relation to, among other indicators, the return on economic investment and the period of emission compensation.

- **Qualitative model.** The evaluation of the social dimension, on the other hand, is exclusively sensitive to the results obtained by means of the surveys carried out and based on the experiences of the end-users of the dwellings. All those results

© The Author(s), under exclusive license to Springer Nature Switzerland AG 2022     43
P. Mercader-Moyano and M. Ramos Martín, *Sustainable Renovation of Buildings*,
SpringerBriefs in Geography, https://doi.org/10.1007/978-3-031-15143-9_4

that are integrated in the IQM to obtain the FCM are part of the set of prerequisites with which the designer must plan the intervention.

For each of the quantitative models, a computer workbook is generated, in which the calculations necessary to obtain the results are programmed and integrated (see CD with appendices) and which are developed in detail in phase 3. The tool developed allows a maximum number of design options (OOD) for each of the envelope regeneration actions:

– EVF-B4 + M1. With a limit of three options at the level of the ventilated façade system (EVF-B4) and a maximum of three options at the level of the insulation to be incorporated (EVF-M1).
– EHCP-B2 + M1. With a limit of two design options for conversion to a ventilated flat roof (EHCP-B2) and a maximum of three options for the insulation to be incorporated (EHCP-M1).
– EVH-C3. With a limit of two design options for the addition of new joinery to the exterior of the opening.

Through the quantitative model in its first phase (IQM), the results of Appendix C have been obtained, as a result of the evaluation of the different strategies that make up the TERP and various DO at product level.

Appendix C tables show environmental and economic quantification selected modules. Based on these design options, a series of combinations of design options (DOCs) are made for each of the renovation action (EVF-B4 + M1, EHCP-B2 + M1 and EVH-C3), which are analysed and ranked according to their impacts from an environmental perspective and from an economic perspective. These design options are as follows:

• **EVF-B4 + M1.** Vertical thermal envelope renovation—façade elements:

  (a) Bioclimatic transformation strategy—transformation to ventilated façade (B4):
      (i)   Option 1. Cladding for ventilated façade formation with decorative high-pressure laminate panel (HPL).
      (ii)  Option 2. Cladding for ventilated façade formation with single-sided polymer concrete parts.
      (iii) Option 3. Cladding for ventilated façade formation with ceramic tiles.
  (b) Thermal insulation improvement strategy—incorporation of insulation on the outside of the façade (M1):
      (i)   Option 1. Rigid glass mineral wool board (MW).
      (ii)  Option 2. Expanded polystyrene board (EPS).
      (iii) Option 3. Extruded polystyrene sheet (XPS).
• **EHCP-B2 + M1.** Horizontal thermal envelope of the flat roof renovation:

  (a) Bioclimatic transformation strategy—transformation to ventilated flat roof (B2):
      (i)   Option 1. Energy rehabilitation with filter slab for formation of a walkable roof with ventilated chamber.

      (ii)  Option 2. Inverted walkable roof with slopes of terrazzo paving on supports.
- (b) Thermal insulation improvement strategy—incorporation of insulation on the outside of the roof (M1):
  - (i) Option 1. Rigid glass mineral wool (MW) board.
  - (ii) Option 2. Expanded polystyrene sheet (EPS).
  - (iii) Option 3. Extruded polystyrene sheet (XPS).
- **EVH-C3.** Vertical thermal envelope renovation—opening elements:

- (a) Strategies for changing the construction system for a new one or combination of the existing one with a new one. Addition of new carpentry to the exterior of the opening (C3):
  - (i) Option 1. Addition of non-plasticized PVC windows and balcony doors.
  - (ii) Option 2. Addition of white lacquered aluminium windows and balconies with thermal break.

## 4.2  Discussion

Contrasting results of both models (IQM and FQM) to determine the effect of some indicators on others, according to intervention specifications, becomes fundamental for a successful application of the methodology that provides value and knowledge to stakeholder. The variables involved in this process are multiple and respond to very diverse interests, which is why the starting method established is based on the essential premise of contrasting each of the dimensions from three perspectives: socio-environmental, socio-economic and economic-environmental.

The discussion of the results obtained from the application focuses on the comparison of the quantitative data calculated at the economic-environmental level (IQM) and the qualitative results determined by means of the surveys carried out. The results for the different economic and environmental indicators resulting from the application are given in Table 4.1.

It should be noted that the methodology developed does not have a reference framework to qualify the regeneration project in relation to the economic and environmental aspects that have been socially contextualized. At present, there are no regulatory thresholds for emissions and/or costs of interventions on buildings that would allow us to judge the adequacy of the results obtained on a local scale.

However, the methodology allows the comparison of different combinations of design options (DOCs) in relation to their impact beyond their implementation, having determined the costs (maintenance, replacement and repair) and benefits (energy savings in the operational phase) beyond the implementation of the actions. The main results of the survey justifying the existence of a FQM are as follows:

- 56% of residents interviewed claim to thermally acclimatize their dwelling in summer or winter by using electrical appliances.
- 53% of residents have changed their windows for others with better characteristics less than ten years ago.

**Table 4.1** Representative results assessment for the economic (COINT, COMA, NPV, EAA, CG, TPB€) and environmental (GWP, TPB CO$_2$ equiv) indicators for the three models (TQM, IQM and FQM)

| 001 DOC | | | | 1 | 2 | 3 | Results deviation (%) | | |
|---|---|---|---|---|---|---|---|---|---|
| | | | | | | | 2 from 1 | 3 from 2 | 3 from 1 |
| | | | | TQM. Results with constant energy savings (no climate change considered) | IQM. Results with variables energy savings (climate change is considered) | FQM. Results with variables energy savings and embody consumption trends based on surveys | | | |
| EVF | B4 + M1 | OP1 | Cladding for ventilated façade formation with HPL high-pressure decorative laminate panel | | | | | | |
| | | OP1 | Rigid glass mineral wool board (MW) | | | | | | |
| EHCP | B2 + M1 | OP1 | Prefabricated parts made of lightened and filtering concrete | | | | | | |
| | | OP1 | Rigid glass mineral wool board (MW) | | | | | | |
| EVH | C3 | OP1 | Addition of non-plasticized PVC joinery | | | | | | |

(continued)

**Table 4.1** (continued)

| | | | 1 | 2 | 3 | Results deviation (%) | | |
|---|---|---|---|---|---|---|---|---|
| | | | TQM. Results with constant energy savings (no climate change considered) | IQM. Results with variables energy savings (climate change is considered) | FQM. Results with variables energy savings and embody consumption trends based on surveys | 2 from 1 | 3 from 2 | 3 from 1 |
| Economic indicators | $CO_{INIT}$ | € | 445,723.41 | 445,723.41 | 396,198.51 | = | −12.50 | −12.50 |
| | $CO_{MA}$ | €/year | 5457.87 | 5457.87 | 4890.40 | = | −11.60 | −11.60 |
| | VAN | € | 497,355.23 | 467,981.27 | 152,876.84 | −6.28 | −206.12 | −225.33 |
| | VAE | € | 19,329.96 | 18,188.32 | 5941.63 | −6.28 | −206.12 | −225.33 |
| | CG | € | 1,254,273.72 | 1,264,863.23 | 1,140,173.99 | 0.84 | −10.94 | −10.01 |
| | TPB€ | years | 17.00 | 18.00 | 35.00 | 5.56 | 48.57 | 51.43 |
| Environmental indicators | GWP | kg $CO_2$ equiv | 341,715.11 | 341,715.11 | 255,544.51 | = | −33.72 | −33.72 |
| | TPB $CO_2$ | years | 7.40 | 8.34 | 11.14 | 11.30 | 25.12 | 33.58 |

(continued)

**Table 4.1** (continued)

| 029 DOC | | | | 1 | 2 | 3 | Results deviation (%) | | |
|---|---|---|---|---|---|---|---|---|---|
| | | | | TQM. Results with constant energy savings (no climate change considered) | IQM. Results with variables energy savings (climate change is considered) | FQM. Results with variables energy savings and embody consumption trends based on surveys | 2 from 1 | 3 from 2 | 3 from 1 |
| | EVF | B4 + M1 | OP1 | Cladding for ventilated façade formation with HPL high-pressure decorative laminate panel | | | | | |
| | | | OP3 | Extruded polystyrene sheet (XPS) | | | | | |
| | EHCP | B2 + M1 | OP1 | Prefabricated lightened and filtering concrete parts | | | | | |
| | | | OP3 | Extruded polystyrene sheet (XPS) | | | | | |
| | EVH | C3 | OP1 | Addition of non-plasticized PVC joinery | | | | | |

(continued)

**Table 4.1** (continued)

| | | | 1 TQM. Results with constant energy savings (no climate change considered) | 2 IQM. Results with variables energy savings (climate change is considered) | 3 FQM. Results with variables energy savings and embody consumption trends based on surveys | Results deviation (%) 2 from 1 | 3 from 2 | 3 from 1 |
|---|---|---|---|---|---|---|---|---|
| Economic indicators | $CO_{INIT}$ | € | 437,632.33 | 437,632.33 | 388,107.43 | = | −12.76 | −12.76 |
| | $CO_{MA}$ | €/year | 5439.64 | 5439.64 | 4872.18 | = | −11.65 | −11.65 |
| | VAN | € | 506,073.29 | 476,699.33 | 161,594.90 | −6.16 | −195.00 | −213.17 |
| | VAE | € | 19,668.79 | 18,527.15 | 6280.47 | −6.16 | −195.00 | −213.17 |
| | CG | € | 1,245,880.35 | 1,256,469.87 | 1,131,780.63 | 0.84 | −11.02 | −10.08 |
| | TPB€ | years | 16.00 | 17.00 | 35.00 | 5.88 | 51.43 | 54.29 |
| Environmental indicators | GWP | kg $CO_2$ equiv | 403,613.30 | 403,613.30 | 317,442.70 | = | −27.15 | −27.15 |
| | TPB $CO_2$ | years | 8.74 | 9.85 | 13.84 | 11.30 | 28.80 | 36.84 |

(continued)

**Table 4.1** (continued)

| | | | | 1 | 2 | 3 | Results deviation (%) | | |
| --- | --- | --- | --- | --- | --- | --- | --- | --- | --- |
| | | | | | | | 2 from 1 | 3 from 2 | 3 from 1 |
| | | | | TQM. Results with constant energy savings (no climate change considered) | IQM. Results with variables energy savings (climate change is considered) | FQM. Results with variables energy savings and embody consumption trends based on surveys | | | |
| 043 DOC | EVF | B4 + M1 | OP2 | Cladding for the formation of a ventilated façade with single-sided polymer concrete parts | | | | | |
| | | | OP1 | Rigid glass mineral wool (MW) slab | | | | | |
| | EHCP | B2 + M1 | OP2 | Walkable inverted roof with slopes of terrazzo paving on supports | | | | | |
| | | | OP1 | Rigid glass mineral wool slab (MW) | | | | | |
| | EVH | C3 | OP1 | Addition of non-plasticized PVC joinery | | | | | |

(continued)

**Table 4.1** (continued)

| | | | 1 | 2 | 3 | Results deviation (%) | | |
|---|---|---|---|---|---|---|---|---|
| | | | TQM. Results with constant energy savings (no climate change considered) | IQM. Results with variables energy savings (climate change is considered) | FQM. Results with variables energy savings and embody consumption trends based on surveys | 2 from 1 | 3 from 2 | 3 from 1 |
| Economic indicators | $CO_{INIT}$ | € | 487,727.15 | 487,727.15 | 438,202.24 | = | −11.30 | −11.30 |
| | $CO_{MA}$ | €/year | 3990.73 | 3990.73 | 3423.27 | = | −16.58 | −16.58 |
| | VAN | € | 505,827.61 | 476,453.65 | 161,349.22 | −6.17 | −195.29 | −213.50 |
| | VAE | € | 19,659.24 | 18,517.61 | 6270.92 | −6.17 | −195.29 | −213.50 |
| | CG | € | 1,157,841.57 | 1,157,841.57 | 1,157,841.57 | 0.00 | 0.00 | 0.00 |
| | $TPB€$ | years | 18.00 | 19.00 | 35.00 | 5.26 | 45.71 | 48.57 |
| Environmental indicators | GWP | kg $CO_2$ equiv | 278,362.78 | 278,362.78 | 192,192.18 | = | −44.84 | −44.84 |
| | TPB $CO_2$ | years | 6.03 | 6.79 | 8.38 | 11.30 | 18.89 | 28.06 |

On the other hand, these two facts lead to a reduction in energy costs derived from air-conditioning consumption, both in the current state of building and in the renovated state, and, on the other hand, to a reduction in the costs of intervention at the level of openings, given that it will be difficult to justify to the users of the dwellings to intervene on elements that are not considered to have been depreciated.

These two premises are introduced in the previous quantification model (IQM), forming a final quantification model (FQM), in which the three design options have been compared:

- The selection of DOC with the lowest impact from the environmental perspective (043 combination 043).
- The selection of DOC with the lowest impact from an economic perspective (029 combination).
- The selection of DOC with the lowest environmental and economic impact, respectively (001 combination).

The results for each of the calculated indicators are determined in Table 4.1. In the following lines, they are discussed and contrasted with the results obtained in the initial quantification model (IQM)—i.e. without the consideration of the social dimension—and in the traditional quantification model (TQM)—i.e. without the consideration of social dimension and with constant emission and consumption savings throughout RSP and without considering the alterations that climate change introduces throughout the LCA.

Table 4.1 shows the calculation results for the economic (COINIT, COMA, NPV, EAA, CG, $TPB_\epsilon$) and environmental (GWP, TPB $CO_2$ equiv) indicators calculated for the three models (TQM, IQM and FQM). The results are contextualized for the three DOCs that have been shown to be most effective from an environmental (043 DOC), economic (029 DOC) and balancing economic and environmental impacts (001 DOC) perspective.

### 4.2.1  Results from a Social-Environmental Perspective

Emissions linked to studied DOCs have an environmental impact whose compensation period is 6.79 years for the most environmentally sustainable combination (043 DOC) and 12.98 years for the combination with the highest environmental impact (102 DOC), both values having been calculated for the IQM.

From a socio-environmental perspective, and if we contextualize the results based on the current culture of the neighbours in relation to the use of energy for the thermal conditioning of dwellings, we observe that these emission compensation periods for the options referred to are extended to 8.38 years (043 DOC) and 18.07 years (102 DOC), increasing the emission compensation period over 23.4% and 39.2%, respectively (Fig. 4.1).

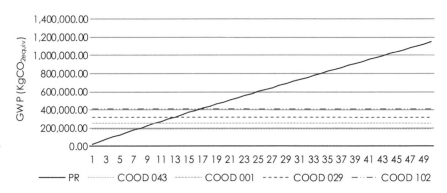

**Fig. 4.1** TPB CO$_2$ for selected design option combination 043, 001, 029 and 102, calculated through FQM

Low participation in the surveys carried out shows the low level of concern generated by raised items. Despite this, it has been possible to confirm the interest of those who have decided to participate, verifying that the building has a large elderly population and that, in practice, they are all owners of the dwellings. The average occupancy of these dwellings is between 2 and 3 people, and from the point of view of adaptability, the degree of satisfaction of the owners with their properties is acceptable. However, the high age profile of the owners will ensure that in less than twenty years, this situation will be completely different, as the owners, their interests in relation to the properties and the quality standards will have changed.

The environmental awareness of the majority of the respondents (92.9%) is verified, and it is confirmed that all respondents would be willing to contribute financially to improve the energy efficiency of their homes and the building as a whole.

The high age profile of the surveyed residents (47.1%) shows that most of the current building conversion needs will change in a period of less than twenty years (when the average life expectancy of the inhabitants is exceeded), which may require conversions from the middle of the remaining lifetime of the buildings to be structured on the basis of the developed methodology. However, it is worth highlighting the interest of the current functional intervention approach in dwellings with an elderly population, especially if we consider the effects of the current pandemic and the tacit need for those most vulnerable to the virus to remain in their dwellings for most of the time. Adapting housing to these needs will entail improving sanitary facilities, recovering outdoor spaces and/or extending them, reducing the number of rooms in favour of creating larger spaces, etc. All of these actions are appropriate for future uses of housing, due to the contribution of value they generate in relation to the new needs that the current crisis has brought to light.

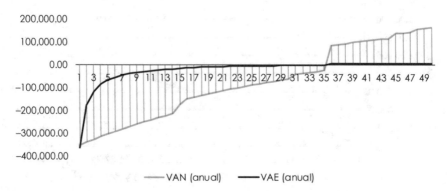

**Fig. 4.2**  Net present value (NPV) and equivalent annual annuity (EAA) throughout RSP to 029 DOC, calculated through FQM

### 4.2.2   Results from a Social-Economic Perspective

It has been verified that the surveyed users are willing to participate financially in the transformation of their buildings to improve their sustainability. The results obtained for economic sustainability assessment were also found to deviate significantly—with an increase of 184.2–205.9% in the periods of financial compensation.

Consideration of aspects such as the percentage of residents who have replaced the window frames with a better one and the low number of users who claim to consume energy for air conditioning throughout the year means that the consideration of the returns on investment determined for 029 DOC—the most economically efficient—is extended from 17 to 35 years (see Fig. 4.2).

Figure 4.2 details, through the NPV and EAA indicators along the RSP, the investment payback period. As can be seen, the reduction in expected consumption over the useful life, which is based on current consumption trends, means that returns on the economic investment are obtained from year 35 onwards, and this time horizon coincides with the expected date on which the useful life of the current envelope system comes to an end.

### 4.2.3   Results from an Environmental-Economic Perspective

In order to draw conclusions on economic and environmental aspects, it is verified that the combination with the lowest economic cost, 029 DOC, has an environmental impact, measured through the GWP indicator, of 317,442.70 kg $CO_2$ equiv, being the sixty-sixth most environmentally sustainable measure, which is far from being a positive aspect from an environmental point of view, also entailing a period of emissions compensation through the expected savings over the useful life of 13.84 years at building level, when considering the consumption trends of users (FQM).

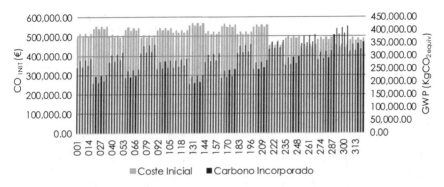

**Fig. 4.3** Initial cost and embodied carbon for each DCO, calculated through FQM

On the other hand, the economic indicators for 043 DOC—lower environmental impact—contextualized at social level, show an overall cost (GC) of 1,157,841.57 € and a payback period of 35 years, which increases the values obtained for the DOC 029 by 2.30% in relation to the GC, but equals in years the payback period. This shows that, at present and for the measures studied, the implementation of measures with a lower environmental impact entails somewhat higher initial investment costs, but these do not represent significant deviations in relation to the periods of return on investment through energy savings.

DOC 001 has the lowest economic and environmental impact of all those studied, with an initial cost of 396,198.51 € and a maintenance cost of 4890.40 € (Fig. 4.3). The emission compensation period is 11.14 years, and the payback period through energy savings equals that of the other options. The consideration of social aspects makes both data deviate from those obtained in the IQM, increasing the TPB by 51.43% and the TPB $CO_2$ by 33.58%.

### 4.2.4  Optimized Thermal Envelope Renovation Project (O-TERP)

Figure 4.4 shows a constructive section of the most sustainable solution from an environmental and economic perspective (001 DOC), which is the Optimized Thermal Envelope Renovation Project (O-TERP). The O-TERP execution will detail the intervention and will have to integrate singularities such as the location of the external air-conditioning units and the layout of telecommunication installations, given that both elements are currently located in façade. For this purpose, as the conclusions go into in greater depth, the conceptualization of a project solution that permits responding to different situations connected with the existing elements in the façade, as well as with the interior redistributions and the openings of new openings or their enlargement that may be required by the users of the dwellings, is a determining factor.

**Fig. 4.4** Type constructive section of DOC 001. (R) Thermal Envelope Regeneration Project (red line): R1, lightweight filtering precast concrete tile; R2, high-pressure decorative laminated panel; R3, mineral wool; R4, insulation reinforcement and R5, windows and balcony doors in non-laminated PVC. (E) Existing thermal envelope: E1, 1/2-foot metric perforated brick; E2, vertical unvented air chamber of 4 cm; E4, plaster, low hardness of 1.5 cm; E5, aluminium window frame; E6, waterproofing sheet and E7, slopes formation and structure

It underlies the importance of considering aspects and dimensions that go beyond the solely consideration of renovation actions at the envelope level that focus exclusively on improving the energy performance of buildings. The fact that these actions add value to the existing dwelling and are capable of improving its internal conditions, comfort and functioning is something that allows us to combine interests with a view to achieving the new European Green Pact.

Consequently, the regeneration of the existing building stock must be based on these objectives and in favour of an improvement in environmental sustainability, although, in the light of the research carried out, these are not the only aspects that can be taken into consideration. In sum, we can state that the improvement of comfort and adaptability are relevant. Actively contributing to the improvement of these aspects through measures to improve energy efficiency is also supported by a change in the current demands on the functionality of dwellings.

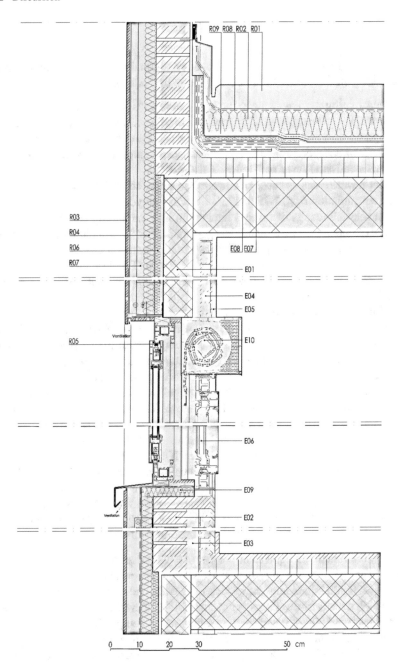

**Fig. 4.4** (continued)

## 4.3  Method Application Conclusions

From a consumption and emissions perspective, thermal model simulation results of
TERP along the Reference Study Period shows the following:

- Effects of climate change along the LCA cause a loss of efficiency of renovation
  actions. We estimate a total consumption and emission reduction at the end of RSP
  compared with initial saves calculated over 17.52% and 17.50%, respectively.
- TERP let reduce total consumption and emissions along the LCA a 21.30%, which
  means a emission saves of 2048.54 TnCO$_2$ equiv.
- Select lower $U$-values cause overheating in future scenarios, doubling refrigera-
  tion consumption from 7.79 to 17.93 kWh/m$^2$ per year.
- Most relevant results of surveys integration into the initial quantification model
  (IQM) let obtain the final quantification model (FQM), concluding the following:
- A lower demand consumption than the initial estimation causes an increase the
  CO$_2$ time payback (TPB CO$_2$) over a 23.4% for DOC 043—the lowest environmental impact combination—and over a 39.2% for DOC 102—the highest
  environmental impact combination.
- Initial cost is lower if we consider individual renovation actions on the thermal
  envelope—mainly windows change and awnings adding. However, economic time
  payback increases because the initial consumption estimation is higher than the
  real ones. We calculate to DOC 029—economic most efficient combination—an
  increase of investment return that goes from 17 to 35 years.
- According to FQM, DOC 001 is the most balance DOC from an environmental
  and economic perspective, with 11.14 years of TPB CO$_2$ and 35 years of TPB.
- To consider climate change along the LCA and to integrate social parameters—
  letting to minimize intervention and to estimate realistic consumption—permit
  detecting remarkable deviations between IQM and FQM. For example, DOC
  001—the most balance intervention strategy—goes for the TPB CO$_2$ indicator
  from 7.40 years (IQM) to 11.14 years (FQM)—increasing a 33.58%—and to
  the TPBEUR indicator from 17 to 35 years—increasing a 51.43%. According to
  these results deviation, we can affirm that climate change and social perspective
  are determinant to get realistic LCA estimations.

From the functional equivalent perspective, we select a representative typology
from a urban scale perspective, considering urban context (1) with a population high
density (200/400 inhabitants per hectare), (2) with a relevant iterative typology, (3)
with a high density of dwellings per hectare (50–100 dwellings), (4) with a medium
net compactly (3.00/4.00) and (5) with a high urban complexity.

According to these premises, we select a urban representative typology from
the Local Agenda XXI of the City (Malaga) and considering refurbishment-interest
areas. This typology is a tower of dwellings with 12 levels and 48 homes.

From this functional equivalent and thanks to an energy performance simulation
through DesignBuilder tool, we define a TERP defined by the following renovation
strategies:

– EVF-B4 + M1. Transformation into ventilated façade + incorporation of insulation on the outside of the façade (0.45 W/m$^2$ K).
– EVH-C3. Addition of new carpentry to the exterior of openings (1.80 W/m$^2$ K).
– EHCP-B2 + M1. Transformation into a ventilated flat roof + incorporation of insulation on the outside of the roof (0.44 W/m$^2$ K).

Method application and indicators calculation let us establish the following conclusions:

- Different simulations are carried out considering climate change on calculations; it is possible to determine throughout the life cycle of renovated building:

  (a) The loss of efficiency of the TERP. This has been quantified through the reduction of 17.52% in total consumption savings and 17.50% in emissions at the end of the Reference Study Period.
  (b) The total reduction in consumption and emissions of the TERP. Verifying a reduction in consumption and emissions of 21.30%, which represents a total emissions saving of 2048.54 TnCO$_2$ equiv.
  (c) The increase in cooling consumption, which increases from 7.79 to 17.93 kWh/m$^2$ year at the end of the TERP's useful life.

- Integration of most relevant results of surveys into the initial quantification model (IQM) let the obtention of a final quantification model (FQM).

  (a) According to TPB CO$_2$, we quantify a deviation of results of FQM compared to IQM caused by a lower consumption demand than initial estimations. It verifies an increment of that period of emission compensation that goes from 23.4 to 39.2% for DOCs.
  (b) Investment return increases, due to mentioned lower consumption demand. We verified that the most economic efficiency design option (029) TPB$_€$ grows a 51.43%.
  (c) Optimal DOC, with economic and environmental impacts more balanced, has a TPB CO$_2$ of 11.14 years and with the year 2055 being the point from which a return on investment is estimated to exist, i.e. having exceeded 70% of the remaining useful life of building.

**Funding** This research has been funded by the research project art. 68/83 LOU ref. 4115/0632 thanks to GRUPO PUMA S.L, ROCKWOOL S.A.U and PROARID GREEN S.L.

# Appendix A

## A.1 Thermal Renovation Strategies Classification and Selection

See Table A.1.

**Table A.1** Thermal renovation strategies classification and selection to energy performance assess on design building

| Thermal renovation strategies | Code | Applies to the case study? | | | |
|---|---|---|---|---|---|
| | | YES | 1 | 2 | 3 |
| Transformation into an extensive/intensive green roof | EHCP-B1 | X | | | |
| Transformation in a ventilated flat roof | EHCP-B2 | X | | | |
| External thermal insulation addition (roof) | EHCP-M1 | X | | | |
| Roof ventilation optimization | EHCI-B1 | | X | | |
| Partial demolition and slope roof substitution | EHCI-C1 | | X | | |
| Partial demolition and change to a flat roof | EHCI-C2 | | X | | |
| Internal thermal insulation addition (floor) | EHS-M1 | | X | | |
| External thermal insulation addition (floor) | EHS-M2 | | X | | |
| Internal thermal insulation addition (ceiling) | EHT-M1 | X | | | |
| Inside thermal insulation addition (under slope roof space) | EHT-M2 | | X | | |
| External greenhouses addition or glass galleries from the outside of the façade | EVF-B1 | | | X | |

(continued)

© The Author(s), under exclusive license to Springer Nature Switzerland AG 2022
P. Mercader-Moyano and M. Ramos Martín, *Sustainable Renovation of Buildings*,
SpringerBriefs in Geography, https://doi.org/10.1007/978-3-031-15143-9

**Table A.1** (continued)

| Thermal renovation strategies | Code | Applies to the case study? | | | |
|---|---|---|---|---|---|
| | | YES | 1 | 2 | 3 |
| Transformation into a thermal collector and accumulator façade | EVF-B2 | | | X | |
| Transformation into a modular green façade | EVF-B3 | | | X | |
| Transformation into a ventilated façade | EVF-B4 | X | | | |
| Transformation into translucent vegetation facade (gallery with vegetation) | EVF-B5 | | X | | |
| Transformation of facade by repairing and painting with thermal insulating paint | EVF-B6 | X | | | |
| Transformation of façade using phase change materials | EVF-B7 | X | | | |
| External thermal insulation addition (façade) | EVF-M1 | X | | | |
| Inside thermal of insulation (façade air chamber) | EVF-M2 | X | | | |
| Internal thermal insulation addition (façade) | EVF-M3 | X | | | |
| External thermal insulation addition (thermal bridge) | EVPT-M1 | X | | | |
| Internal thermal insulation addition (thermal bridge) | EVPT-M2 | | | | X |
| Addition of exterior shading element to the opening | EVH-B1 | X | | | |
| Addition of an interior shading element to the opening | EVH-B2 | X | | | |
| Incorporation of shading elements by means of vegetation | EVH-B3 | X | | | |
| External thermal insulation addition (opening contours) | EVH-M1 | | X | | X |
| Replacement of carpentry | EVH-C1 | X | | | |
| Replacement with low-emissivity glazing | EVH-C2 | X | | | |
| Addition of new carpentry to the exterior of the opening | EVH-C3 | X | | | |
| Incorporation of planters in the opening | EVH-C4 | | X | X | X |
| Transformation into a vertical garden | EVM-B1 | | X | | |
| External thermal insulation addition (party wall) | EVM-M1 | | X | | |

1 No, constructive adaptation to the case study is impossible
2 No, it is necessary to design a non-standardized solution that is readily available in industry
3 No, significant impact on the neighbourhood during implementation of the measure

## A.2 Assessment of the Energy Performance of the Various Thermal Envelope Strategies Classified and Selected

### [B] Bioclimatic Transformation

- **EHCP-B1. Transformation into an extensive/intensive green roof**

See Table A.2.

**Table A.2** Results of assessment of EHCP-B1 strategy and percentage difference from the current status

| EHCP-B1 | | | Zone A | Zone B | Zone C | Zone D |
|---|---|---|---|---|---|---|
| | | | N–O | N–E | S–E | S–O |
| P12 | Comfort | Difference of operative temperature from the current status (°C) | −0.12 | −0.12 | −0.13 | −0.12 |
| | | Difference of humidity from the current status (%) | 0.29 | 0.32 | 0.32 | 0.27 |
| | Solar gains (Wh/m$^2$) | | 0 | 0 | 0 | 0 |
| | Percentage difference from the current status | | 0 | 0 | 0 | 0 |
| | Sensible heating (Wh/m$^2$) | | −2261.48 | −1938.09 | −2057.67 | −2292.64 |
| | Percentage difference from the current status (%) | | −6.62 | −5.17 | −5.88 | −7.17 |
| | Sensible cooling (Wh/m$^2$) | | 3113.26 | 2941.66 | 2919.03 | 3152.67 |
| | Percentage difference from the current status (%) | | −15.78 | −17.84 | −17.48 | −15.21 |

- **EHCP-B2. Transformation in a ventilated flat roof**

See Table A.3.

**Table A.3** Results of assessment of EHCP-B2 strategy and percentage difference from the current status

| EHCP-B2 | | | Zone A | Zone B | Zone C | Zone D |
|---|---|---|---|---|---|---|
| | | | N–O | N–E | S–E | S–O |
| P12 | Comfort | Difference of operative temperature from the current status (°C) | −0.18 | −0.18 | −0.19 | −0.17 |
| | | Difference of humidity from the current status (%) | 0.43 | 0.46 | 0.45 | 0.41 |
| | Solar gains (Wh/m$^2$) | | | 0 | 0 | 0 |
| | Percentage difference from the current status | | | 0 | 0 | 0 |
| | Sensible heating (Wh/m$^2$) | | | −1298.77 | −1449.83 | −1723.44 |
| | Percentage difference from the current status (%) | | | −3.47 | −4.14 | −5.39 |
| | Sensible cooling (Wh/m$^2$) | | | 3396.1 | 3379.17 | 3644.87 |
| | Percentage difference from the current status (%) | | | −20.59 | −20.24 | −17.59 |

- **EVF-B4. Transformation into a ventilated façade**

See Table A.4.

**Table A.4** Results of assessment of EHCP-B4 strategy and percentage difference from the current status

| EVF-B4 | | | Zone A | Zone B | Zone C | Zone D |
|---|---|---|---|---|---|---|
| | | | N–O | N–E | S–E | S–O |
| P01 | Comfort | Difference of operative temperature from the current status (°C) | 0.05 | 0.05 | 0.02 | 0.01 |
| | | Difference of humidity from the current status (%) | −0.11 | −0.12 | −0.05 | −0.03 |
| | Solar gains (Wh/m$^2$) | | | 0 | 0 | 0 |
| | Percentage difference from the current status | | | 0 | 0 | 0 |
| | Sensible heating (Wh/m$^2$) | | | −2408.18 | −1483.31 | −1303.15 |
| | Percentage difference from the current status (%) | | | −10.17 | −8.87 | −9.11 |
| | Sensible cooling (Wh/m$^2$) | | | 821.44 | 1110.96 | 1447.43 |
| | Percentage difference from the current status (%) | | | −6.42 | −8.18 | −8.45 |
| P02–P11 | Comfort | Difference of operative temperature from the current status (°C) | 0.06 | 0.07 | 0.04 | 0.04 |
| | | Difference of humidity from the current status (%) | −0.14 | −0.16 | −0.11 | −0.08 |
| | Solar gains (Wh/m$^2$) | | | 0 | 0 | 0 |
| | Percentage difference from the current status | | | 0 | 0 | 0 |
| | Sensible heating (Wh/m$^2$) | | | −2718.56 | −1859.21 | −1653.98 |
| | Percentage difference from the current status (%) | | | −10.57 | −9.77 | −10.13 |
| | Sensible cooling (Wh/m$^2$) | | | 656.19 | 901.72 | 1223.82 |
| | Percentage difference from the current status (%) | | | −5.60 | −7.30 | −7.75 |
| P12 | Comfort | Difference of operative temperature from the current status (°C) | 0.03 | 0.04 | 0.03 | 0.01 |
| | | Difference of humidity from the current status (%) | −0.05 | −0.08 | −0.06 | −0.04 |
| | Solar gains (Wh/m$^2$) | | | 0 | 0 | 0 |
| | Percentage difference from the current status | | | 0 | 0 | 0 |
| | Sensible heating (Wh/m$^2$) | | | −2088.89 | −1847.71 | −1751.68 |
| | Percentage difference from the current status (%) | | | −5.58 | −5.28 | −5.48 |
| | Sensible cooling (Wh/m$^2$) | | | 802.27 | 870.88 | 1234.93 |
| | Percentage difference from the current status (%) | | | −4.87 | −5.22 | −5.96 |

- **EVF-B6. Transformation of facade by repairing and painting with thermal insulating paint**

See Table A.5.

**Table A.5** Results of assessment of EVF-B6 strategy and percentage difference from the current status

| EVF-B6 | | | Zone A | Zone B | Zone C | Zone D |
|---|---|---|---|---|---|---|
| | | | N–O | N–E | S–E | S–O |
| P01 | Comfort | Difference of operative temperature from the current status (°C) | 0.08 | 0.09 | 0.05 | 0.05 |
| | | Difference of humidity from the current status (%) | −0.18 | −0.19 | −0.12 | −0.1 |
| | Solar gains (Wh/m$^2$) | | | 0 | 0 | 0 |
| | Percentage difference from the current status | | | 0 | 0 | 0 |
| | Sensible heating (Wh/m$^2$) | | | −3086.38 | −2012.04 | −1796.38 |
| | Percentage difference from the current status (%) | | | −13.04 | −12.03 | −12.56 |
| | Sensible cooling (Wh/m$^2$) | | | 812.7 | 134.8 | 1455.97 |
| | Percentage difference from the current status (%) | | | −6.35 | −0.99 | −8.50 |
| P02–P11 | Comfort | Difference of operative temperature from the current status (°C) | 0.09 | 0.07 | 0.07 | 0.07 |
| | | Difference of humidity from the current status (%) | −0.21 | −0.22 | −0.17 | −0.16 |
| | Solar gains (Wh/m$^2$) | | | 0 | 0 | 0 |
| | Percentage difference from the current status | | | 0 | 0 | 0 |
| | Sensible heating (Wh/m$^2$) | | | −3414.73 | −2426.59 | −2195.34 |
| | Percentage difference from the current status (%) | | | −13.28 | −12.76 | −13.44 |
| | Sensible cooling (Wh/m$^2$) | | | 627 | 901.79 | 1208.64 |
| | Percentage difference from the current status (%) | | | −5.35 | −7.30 | −7.65 |
| P12 | Comfort | Difference of operative temperature from the current status (°C) | 0.05 | 0.06 | 0.05 | 0.03 |
| | | Difference of humidity from the current status (%) | −0.09 | −0.12 | −0.1 | −0.07 |
| | Solar gains (Wh/m$^2$) | | | 0 | 0 | 0 |
| | Percentage difference from the current status | | | 0 | 0 | 0 |
| | Sensible heating (Wh/m$^2$) | | | −2674.04 | −2393.65 | −2310.65 |
| | Percentage difference from the current status (%) | | | −7.14 | −6.84 | −7.23 |
| | Sensible cooling (Wh/m$^2$) | | | 769.48 | 844.96 | 1197.5 |
| | Percentage difference from the current status (%) | | | −4.67 | −5.06 | −5.78 |

- **EVF-B7. Transformation of façade using phase change materials**

The solution presented here involves the treatment of the entire exterior surface of the building with an amorphous material which, after its application, offers theoretical transmittance values of 0.753 W/m$^2$ K, taking into account thermal bridges.

In order to define the phase change behaviour of the material, four points have been defined in the temperature–enthalpy curve: Point 1, for a temperature of −20 °C, an enthalpy value of 0 J/kg is established; Point 2, the value of 20 °C is related to an enthalpy of 33,400 J/kg; Point 3, for a temperature of 20.50 °C, a value of 70,000 J/kg is set, and Point 4, for a temperature of 100.00 °C, an enthalpy of 137,000 J/kg is established. In the same way, a specific calculation algorithm based on finite differences has been defined in the programme to simulate the behaviour of the material depending on the outside climate.

This measure leads to significant improvements in the winter months and not so much in the summer months, partly due to the fact that no heat is dissipated from the inside of the house to the outside through the envelope (see Table A.6).

**Table A.6** Results of assessment of EVF-B7 strategy and percentage difference from the current status

| EVF-B7 | | | Zone A | Zone B | Zone C | Zone D |
|---|---|---|---|---|---|---|
| | | | N–O | N–E | S–E | S–O |
| P01 | Comfort | Difference of operative temperature from the current status (°C) | 0.43 | 0.42 | 0.38 | 0.38 |
| | | Difference of humidity from the current status (%) | −0.96 | −0.93 | −0.87 | −0.88 |
| | Solar gains (Wh/m$^2$) | | | 7.46 | 10.22 | −7.23 |
| | Percentage difference from the current status (%) | | | 0.02 | 0.03 | −0.02 |
| | Sensible heating (Wh/m$^2$) | | | −8805.25 | −6420.8 | −5733.71 |
| | Percentage difference from the current status (%) | | | −37.19 | −38.37 | −40.08 |
| | Sensible cooling (Wh/m$^2$) | | | 326.42 | 864.98 | 1300.81 |
| | Percentage difference from the current status (%) | | | −2.55 | −6.37 | −7.59 |
| P02–P11 | Comfort | Difference of operative temperature from the current status (°C) | 0.44 | 0.43 | 0.4 | 0.41 |
| | | Difference of humidity from the current status (%) | −0.98 | −0.96 | −0.93 | −0.94 |
| | Solar gains (Wh/m$^2$) | | | 10.29 | 10.21 | −7.08 |
| | Percentage difference from the current status (%) | | | 0.03 | 0.03 | −0.02 |
| | Sensible heating (Wh/m$^2$) | | | −9515.9 | −7344.56 | 81,377.9 |
| | Percentage difference from the current status (%) | | | −37.00 | −38.61 | 498.21 |
| | Sensible cooling (Wh/m$^2$) | | | 50.95 | 507.23 | 910.14 |
| | Percentage difference from the current status (%) | | | −0.43 | −4.10 | −5.76 |

(continued)

**Table A.6**  (continued)

| EVF-B7 | | | Zone A | Zone B | Zone C | Zone D |
|---|---|---|---|---|---|---|
| | | | N–O | N–E | S–E | S–O |
| P12 | Comfort | Difference of operative temperature from the current status (°C) | 0.23 | 0.25 | 0.23 | 0.21 |
| | | Difference of humidity from the current status (%) | −0.44 | −0.49 | −0.46 | −0.42 |
| | Solar gains (Wh/m²) | | | 20.97 | 25.06 | 3.77 |
| | Percentage difference from the current status (%) | | | 0.05 | 0.06 | 0.01 |
| | Sensible heating (Wh/m²) | | | −7478.95 | −6924.59 | −6601.51 |
| | Percentage difference from the current status (%) | | | −19.96 | −19.79 | −20.64 |
| | Sensible cooling (Wh/m²) | | | 85.09 | 215.57 | 731.83 |
| | Percentage difference from the current status (%) | | | −0.52 | −1.29 | −3.53 |

- **EVH-B1. Addition of exterior shading element to the opening (façade level)**

  See Table A.7.

**Table A.7**  Results of assessment of EVH-B1 strategy and percentage difference from the current status

| EVH-B1 | | | Zone A | Zone B | Zone C | Zone D |
|---|---|---|---|---|---|---|
| | | | N–O | N–E | S–E | S–O |
| P01 | Comfort | Difference of operative temperature from the current status (°C) | −0.23 | −0.17 | −0.27 | −0.37 |
| | | Difference of humidity from the current status (%) | 0.54 | 0.41 | 0.65 | 0.88 |
| | Solar gains (Wh/m²) | | | −4527.52 | −5698.07 | −8229.87 |
| | Percentage difference from the current status (%) | | | −13.48 | −14.71 | −17.74 |
| | Sensible heating (Wh/m²) | | | 1018.39 | 1458.91 | 1867.08 |
| | Percentage difference from the current status (%) | | | 4.30 | 8.72 | 13.05 |
| | Sensible cooling (Wh/m²) | | | 1520.38 | 2327.9 | 3778.34 |
| | Percentage difference from the current status (%) | | | −11.89 | −17.15 | −22.05 |
| P02– P11 | Comfort | Difference of operative temperature from the current status (°C) | −0.22 | −0.16 | −0.25 | −0.33 |
| | | Difference of humidity from the current status (%) | 0.51 | 0.37 | 0.59 | 0.81 |
| | Solar gains (Wh/m²) | | | −4511.71 | −5674.19 | −8215.38 |
| | Percentage difference from the current status (%) | | | −13.41 | −14.65 | −17.70 |
| | Sensible heating (Wh/m²) | | | 1055.29 | 1489.81 | 1931.23 |
| | Percentage difference from the current status (%) | | | 4.10 | 7.83 | 11.82 |
| | Sensible cooling (Wh/m²) | | | 1385.55 | 2075.81 | 3433.72 |
| | Percentage difference from the current status (%) | | | −11.83 | −16.80 | −21.75 |

(continued)

**Table A.7** (continued)

| EVH-B1 | | | Zone A | Zone B | Zone C | Zone D |
|---|---|---|---|---|---|---|
| | | | N–O | N–E | S–E | S–O |
| P12 | Comfort | Difference of operative temperature from the current status (°C) | −0.11 | −0.07 | −0.16 | −0.21 |
| | | Difference of humidity from the current status (%) | 0.25 | 0.18 | 0.36 | 0.47 |
| | Solar gains (Wh/m$^2$) | | | −3571.7 | −5473.87 | −7257.28 |
| | Percentage difference from the current status (%) | | | −8.73 | −12.65 | −14.01 |
| | Sensible heating (Wh/m$^2$) | | | 893.69 | 1408.11 | 1857.73 |
| | Percentage difference from the current status (%) | | | 2.39 | 4.02 | 5.81 |
| | Sensible cooling (Wh/m$^2$) | | | 832.68 | 1798.21 | 2746.32 |
| | Percentage difference from the current status (%) | | | −5.05 | −10.77 | −13.25 |

- **EVH-B2. Addition of an interior shading element to the opening**

  See Table A.8.

**Table A.8** Results of assessment of EVH-B2 strategy and percentage difference from the current status

| EVH-B2 | | | Zone A | Zone B | Zone C | Zone D |
|---|---|---|---|---|---|---|
| | | | N–O | N–E | S–E | S–O |
| P01 | Comfort | Difference of operative temperature from the current status (°C) | −0.32 | −0.26 | −0.32 | −0.41 |
| | | Difference of humidity from the current status (%) | 0.73 | 0.59 | 0.69 | 0.9 |
| | Solar gains (Wh/m$^2$) | | | −24,130.38 | −28,524.72 | −34,955.99 |
| | Percentage difference from the current status (%) | | | −71.83 | −73.66 | −75.36 |
| | Sensible heating (Wh/m$^2$) | | | 972.38 | 1642.36 | 2174.51 |
| | Percentage difference from the current status (%) | | | 4.11 | 9.82 | 15.20 |
| | Sensible cooling (Wh/m$^2$) | | | 2689.02 | 2655.32 | 2410.62 |
| | Percentage difference from the current status (%) | | | −21.02 | −19.56 | −14.07 |
| P02–P11 | Comfort | Difference of operative temperature from the current status (°C) | −0.32 | −0.26 | −0.31 | −0.4 |
| | | Difference of humidity from the current status (%) | 0.73 | 0.59 | 0.69 | 0.89 |
| | Solar gains (Wh/m$^2$) | | | −24,182.18 | −28,537.64 | −34,971.25 |
| | Percentage difference from the current status (%) | | | −71.85 | −73.66 | −75.36 |
| | Sensible heating (Wh/m$^2$) | | | 975.95 | 1691.93 | 2351.39 |
| | Percentage difference from the current status (%) | | | 3.79 | 8.89 | 14.40 |

(continued)

**Table A.8** (continued)

| EVH-B2 | | | Zone A | Zone B | Zone C | Zone D |
|---|---|---|---|---|---|---|
| | | | N–O | N–E | S–E | S–O |
| | Sensible cooling (Wh/m$^2$) | | | 2623.91 | 2585.02 | 2397.8 |
| | Percentage difference from the current status (%) | | | −22.40 | −20.92 | −15.19 |
| P12 | Comfort | Difference of operative temperature from the current status (°C) | −0.26 | −0.2 | −0.22 | −0.3 |
| | | Difference of humidity from the current status (%) | 0.55 | 0.44 | 0.46 | 0.6 |
| | Solar gains (Wh/m$^2$) | | | −30,380.28 | −32,378.41 | −39,557.81 |
| | Percentage difference from the current status (%) | | | −74.23 | −74.85 | −76.37 |
| | Sensible heating (Wh/m$^2$) | | | 1912.65 | 2163.88 | 2951.5 |
| | Percentage difference from the current status (%) | | | 5.10 | 6.18 | 9.23 |
| | Sensible cooling (Wh/m$^2$) | | | 2561.72 | 2565.22 | 2324.25 |
| | Percentage difference from the current status (%) | | | −15.53 | −15.37 | −11.22 |

- **EVH-B3. Addition of exterior shading element to the opening (window level)**

See Table A.9.

**Table A.9** Results of assessment of EVH-B3 strategy and percentage difference from the current status

| EVH-B3 | | | Zone A | Zone B | Zone C | Zone D |
|---|---|---|---|---|---|---|
| | | | N–O | N–E | S–E | S–O |
| P01 | Comfort | Difference of operative temperature from the current status (°C) | −0.21 | −0.17 | −0.26 | −0.33 |
| | | Difference of humidity from the current status (%) | 0.5 | 0.41 | 0.62 | 0.78 |
| | Solar gains (Wh/m$^2$) | | | −5953.38 | −8864.16 | −11,816.36 |
| | Percentage difference from the current status (%) | | | −17.72 | −22.89 | −25.47 |
| | Sensible heating (Wh/m$^2$) | | | 841.92 | 2051.92 | 2127.12 |
| | Percentage difference from the current status (%) | | | 3.56 | 12.26 | 14.87 |
| | Sensible cooling (Wh/m$^2$) | | | 1525.6 | 1771.57 | 3152.16 |
| | Percentage difference from the current status (%) | | | −11.93 | −13.05 | −18.39 |

(continued)

**Table A.9** (continued)

| EVH-B3 | | | Zone A | Zone B | Zone C | Zone D |
|---|---|---|---|---|---|---|
| | | | N–O | N–E | S–E | S–O |
| P02–P11 | Comfort | Difference of operative temperature from the current status (°C) | −0.21 | −0.17 | −0.24 | −0.31 |
| | | Difference of humidity from the current status (%) | 0.5 | 0.39 | 0.58 | 0.74 |
| | Solar gains (Wh/m²) | | | −6009.91 | −8879.22 | −11,833.7 |
| | Percentage difference from the current status (%) | | | −17.86 | −22.92 | −25.50 |
| | Sensible heating (Wh/m²) | | | 905.2 | 2137.16 | 2287.93 |
| | Percentage difference from the current status (%) | | | 3.52 | 11.23 | 14.01 |
| | Sensible cooling (Wh/m²) | | | 1452.05 | 1653.83 | 2971.03 |
| | Percentage difference from the current status (%) | | | −12.39 | −13.38 | −18.82 |
| P12 | Comfort | Difference of operative temperature from the current status (°C) | −0.23 | −0.18 | −0.21 | −0.27 |
| | | Difference of humidity from the current status (%) | 0.53 | 0.43 | 0.49 | 0.6 |
| | Solar gains (Wh/m²) | | | −11,013.26 | −12,229.31 | −15,761.31 |
| | Percentage difference from the current status (%) | | | −26.91 | −28.27 | −30.43 |
| | Sensible heating (Wh/m²) | | | 2330.3 | 2551.87 | 2946.83 |
| | Percentage difference from the current status (%) | | | 6.22 | 7.29 | 9.21 |
| | Sensible cooling (Wh/m²) | | | 1851.67 | 2042.96 | 3667.75 |
| | Percentage difference from the current status (%) | | | −11.23 | −12.24 | −17.70 |

- **Bioclimatic transformation comparison**

  See Table A.10.

**Table A.10** Results of bioclimatic transformation strategies assessment and comparison

| | | | EA | [B] Bioclimatic transformation | | | | | | | |
| | | | | EHCP | | EVF | | | EVH | | |
| | | | | B1 | B2 | B4 | B6 | B7 | B1 | B2 | B3 |
|---|---|---|---|---|---|---|---|---|---|---|---|
| Comfort | Temperature (°C) | Air temperature | 22.04 | 22.03 | 22.03 | 22.07 | 22.09 | 22.33 | 21.87 | 21.82 | 21.87 |
| | | Radiant temperature | 22.31 | 22.3 | 22.29 | 22.37 | 22.42 | 22.82 | 22.02 | 21.9 | 22.02 |
| | | Operative temperature | 22.17 | 22.17 | 22.16 | 22.22 | 22.25 | 22.57 | 21.95 | 21.86 | 21.95 |
| | Relative humidity (%) | | 53.72 | 53.74 | 53.75 | 53.61 | 53.54 | 52.82 | 54.27 | 54.44 | 54.27 |
| Indoor gains (Wh/m²) | Solar gains (windows) | | 37,147 | 37,147 | 37,147 | 37,147 | 37,147 | 37,156 | 31,505 | 9684 | 28,747 |
| | Sensible heating of zone | | 21,043 | 20,877 | 20,925 | 19,023 | 18,434 | 13,483 | 22,501 | 22,821 | 22,746 |
| | Sensible cooling of zone | | -13,640 | -13,396 | -13,357 | -12,765 | -12,790 | -13,242 | -11,477 | -11,235 | -11,235 |
| Emissions (kg CO₂ equiv) | CO₂ production | | 35,574 | 35,390 | 35,411 | 33,978 | 33,598 | 30,479 | 35,868 | 36,003 | 36,054 |
| | Heating percentage of savings from the current status (%) | | | 0.5 | 0.5 | 4.5 | 5.6 | 14.3 | -0.8 | -1.2 | -1.3 |
| Consumptions (Wh/m²) | Heating consumptions savings from the current status | | 22,873 | 22,692 | 22,745 | 20,677 | 20,037 | 14,655 | 24,458 | 24,805 | 24,724 |
| | Cooling percentage of savings from the current status (%) | | | 0.8 | 0.6 | 9.6 | 12.4 | 35.9 | -6.9 | -8.4 | -8.1 |
| | Cooling consumptions savings from the current status | | 6922 | 6800 | 6780 | 6484 | 6498 | 6732 | 5823 | 5699 | 5863 |
| | Total HVAC percentage of savings from the current status (%) | | | 1.8 | 2.0 | 6.3 | 6.1 | 2.7 | 15.9 | 17.7 | 15.3 |

## [C] Changes/Combination by/with New Construction Systems

- **EVH-C1. Replacement of carpentry**

    See Table A.11.

**Table A.11** Results of assessment of EVH-C1 strategy and percentage difference from the current status

| EVH-C1 | | | Zone A | Zone B | Zone C | Zone D |
|---|---|---|---|---|---|---|
| | | | N–O | N–E | S–E | S–O |
| P01 | Comfort | Difference of operative temperature from the current status (°C) | 0.36 | 0.34 | 0.39 | 0.41 |
| | | Difference of humidity from the current status (%) | −0.87 | −1.83 | −0.97 | −1.01 |
| | Solar gains (Wh/m²) | | | 3800.17 | 4246.32 | 5275.33 |
| | Percentage difference from the current status (%) | | | 11.31 | 10.97 | 11.37 |
| | Sensible heating (Wh/m²) | | | −5170.17 | −4692.19 | −4127.36 |
| | Percentage difference from the current status (%) | | | −21.84 | −28.04 | −28.85 |
| | Sensible cooling (Wh/m²) | | | −1087.34 | −1036.51 | −942.54 |
| | Percentage difference from the current status (%) | | | 8.50 | 7.63 | 5.50 |
| P02–P11 | Comfort | Difference of operative temperature from the current status (°C) | 0.36 | 0.35 | 0.4 | 0.42 |
| | | Difference of humidity from the current status (%) | −0.89 | −0.87 | −1 | −1.03 |
| | Solar gains (Wh/m²) | | | 3798.3 | 4247.98 | 5277.3 |
| | Percentage difference from the current status (%) | | | 11.29 | 10.97 | 11.37 |
| | Sensible heating (Wh/m²) | | | −5841.23 | −5449.29 | −4787.4 |
| | Percentage difference from the current status (%) | | | −22.71 | −28.64 | −29.31 |
| | Sensible cooling (Wh/m²) | | | −1187.75 | −1117.53 | −1068.7 |
| | Percentage difference from the current status (%) | | | 10.14 | 9.04 | 6.77 |
| P12 | Comfort | Difference of operative temperature from the current status (°C) | 0.26 | 0.27 | 0.27 | 0.27 |
| | | Difference of humidity from the current status (%) | −0.62 | −0.63 | −0.65 | −0.64 |
| | Solar gains (Wh/m²) | | | 4453.13 | 4720.35 | 5829.95 |
| | Percentage difference from the current status (%) | | | 10.88 | 10.91 | 11.26 |
| | Sensible heating (Wh/m²) | | | −5636 | −5600.09 | −5193.19 |
| | Percentage difference from the current status (%) | | | −15.04 | −16.00 | −16.24 |
| | Sensible cooling (Wh/m²) | | | −1265.1 | −1231.48 | −1122.99 |
| | Percentage difference from the current status (%) | | | 7.67 | 7.38 | 5.42 |

- **EVH-C2. Replacement with low-emissivity glazing**

See Table A.12.

**Table A.12** Results of assessment of EVH-C2 strategy and percentage difference from the current status

| EVH-C2 | | | Zone A | Zone B | Zone C | Zone D |
|---|---|---|---|---|---|---|
| | | | N–O | N–E | S–E | S–O |
| P01 | Comfort | Difference of operative temperature from the current status (°C) | −0.02 | 0 | 0 | −0.04 |
| | | Difference of humidity from the current status (%) | 0.01 | −0.04 | −0.04 | 0.05 |
| | Solar gains (Wh/m$^2$) | | | −11,685.49 | −13,639.44 | −16,151.85 |
| | Percentage difference from the current status (%) | | | −34.78 | −35.22 | −34.82 |
| | Sensible heating (Wh/m$^2$) | | | −2826.66 | −2042.65 | −1547.85 |
| | Percentage difference from the current status (%) | | | −11.94 | −12.21 | −10.82 |
| | Sensible cooling (Wh/m$^2$) | | | 1482.65 | 1498.42 | 2159.51 |
| | Percentage difference from the current status (%) | | | −11.59 | −11.04 | −12.60 |
| P02–P11 | Comfort | Difference of operative temperature from the current status (°C) | −0.02 | 0.01 | 0.01 | −0.03 |
| | | Difference of humidity from the current status (%) | 0 | −0.06 | −0.06 | 0.03 |
| | Solar gains (Wh/m$^2$) | | | −11,713.37 | −13,644.81 | −16,158.19 |
| | Percentage difference from the current status (%) | | | −34.80 | −35.22 | −34.82 |
| | Sensible heating (Wh/m$^2$) | | | −3267.57 | −2451.54 | −1852.51 |
| | Percentage difference from the current status (%) | | | −12.70 | −12.89 | −11.34 |
| | Sensible cooling (Wh/m$^2$) | | | 1416.44 | 1428.69 | 2060.48 |
| | Percentage difference from the current status (%) | | | −12.09 | −11.56 | −13.05 |
| P12 | Comfort | Difference of operative temperature from the current status (°C) | −0.02 | 0.01 | 0 | −0.04 |
| | | Difference of humidity from the current status (%) | 0.02 | −0.04 | −0.03 | 0.05 |
| | Solar gains (Wh/m$^2$) | | | −14,436.06 | −15,264.43 | −18,107.96 |
| | Percentage difference from the current status (%) | | | −35.27 | −35.29 | −34.96 |
| | Sensible heating (Wh/m$^2$) | | | −2536.58 | −2253.22 | −1621.99 |
| | Percentage difference from the current status (%) | | | −6.77 | −6.44 | −5.07 |
| | Sensible cooling (Wh/m$^2$) | | | 1306.9 | 1321.7 | 2099.62 |
| | Percentage difference from the current status (%) | | | −7.93 | −7.92 | −10.13 |

- **EVH-C3. Addition of new carpentry to the exterior of the opening**

  See Table A.13.

**Table A.13** Results of assessment of EVH-C3 strategy and percentage difference from the current status

| EVH-C3 | | | Zone A | Zone B | Zone C | Zone D |
|---|---|---|---|---|---|---|
| | | | N–O | N–E | S–E | S–O |
| P01 | Comfort | Difference of operative temperature from the current status (°C) | −0.09 | −0.06 | −0.08 | −0.14 |
| | | Difference of humidity from the current status (%) | 0.19 | 0.11 | 0.13 | 0.27 |
| | Solar gains (Wh/m²) | | | −17,776.5 | −20,652.42 | −24,551.3 |
| | Percentage difference from the current status (%) | | | −52.91 | −53.33 | −52.93 |
| | Sensible heating (Wh/m²) | | | −3060.03 | −1990.53 | −1369.98 |
| | Percentage difference from the current status (%) | | | −12.93 | −11.90 | −9.58 |
| | Sensible cooling (Wh/m²) | | | 2353.43 | 2369.71 | 3281.39 |
| | Percentage difference from the current status (%) | | | −18.40 | −17.45 | −19.15 |
| P02–P11 | Comfort | Difference of operative temperature from the current status (°C) | −0.09 | −0.05 | −0.07 | −0.12 |
| | | Difference of humidity from the current status (%) | 0.17 | 0.08 | 0.1 | 0.24 |
| | Solar gains (Wh/m²) | | | −17,815.08 | −20,660.55 | −24,560.91 |
| | Percentage difference from the current status (%) | | | −52.94 | −53.33 | −52.93 |
| | Sensible heating (Wh/m²) | | | −3553.71 | −2438.63 | −1674.2 |
| | Percentage difference from the current status (%) | | | −13.82 | −12.82 | −10.25 |
| | Sensible cooling (Wh/m²) | | | 2245.97 | 2255.77 | 3126.92 |
| | Percentage difference from the current status (%) | | | −19.17 | −18.25 | −19.80 |
| P12 | Comfort | Difference of operative temperature from the current status (°C) | −0.08 | −0.04 | −0.05 | −0.11 |
| | | Difference of humidity from the current status (%) | 0.15 | 0.07 | 0.08 | 0.19 |
| | Solar gains (Wh/m²) | | | −21,848.23 | −23,096.29 | −27,484.41 |
| | Percentage difference from the current status (%) | | | −53.38 | −53.39 | −53.06 |
| | Sensible heating (Wh/m²) | | | −2539.4 | −2136.2 | −1293.74 |
| | Percentage difference from the current status (%) | | | −6.78 | −6.10 | −4.05 |
| | Sensible cooling (Wh/m²) | | | 2085.62 | 2104.37 | 3174.42 |
| | Percentage difference from the current status (%) | | | −12.65 | −12.61 | −15.32 |

- **Changes/combination by/with new construction systems comparison**

See Table A.14.

**Table A.14** Results of changes/combination by/with new construction systems strategies assessment

| | | | EA | [C] Changes/combination by/with new construction systems | | |
|---|---|---|---|---|---|---|
| | | | | EVH | | |
| | | | | C1 | C2 | C3 |
| Comfort | Temperature (°C) | Air temperature | 22.04 | 22.32 | 22.04 | 21.99 |
| | | Radiant temperature | 22.31 | 22.76 | 22.29 | 22.2 |
| | | Operative temperature | 22.17 | 22.57 | 22.16 | 22.09 |
| | Relative humidity (%) | | 53.72 | 52.82 | 53.71 | 53.88 |
| Indoor gains (Wh/m²) | Solar gains (windows) | | 37,147 | 41,333 | 24,186 | 17,457 |
| | Sensible heating of zone | | 21,043 | 13,483 | 18,740 | 18,751 |
| | Sensible cooling of zone | | −13,640 | −13,242 | −12,010 | −11,114 |
| Emissions (kg CO₂ equiv) | $CO_2$ production | | 35,574 | 32,550 | 33,555 | 33,286 |
| | Heating percentage of savings from the current status (%) | | | 8.5 | 5.7 | 6.4 |
| Consumptions (Wh/m²) | Heating consumptions savings from the current status | | 22,873 | 14,655 | 20,370 | 20,382 |
| | Cooling percentage of savings from the current status (%) | | | 35.9 | 10.9 | 10.9 |
| | Cooling consumptions savings from the current status | | 6922 | 6732 | 6094 | 5638 |
| | Total HVAC percentage of savings from the current status (%) | | | 2.7 | 12.0 | 18.5 |

## *[M] Improvement of Insulation*

- **EHCP-M1. External thermal insulation addition (roof)**

  See Table A.15.

**Table A.15** Results of assessment of EHCP-M1 strategy and percentage difference from the current status

| EHCP-M1 | | | Zone A | Zone B | Zone C | Zone D |
|---------|---|---|--------|--------|--------|--------|
| | | | N–O | N–E | S–E | S–O |
| P12 | Comfort | Difference of operative temperature from the current status (°C) | −0.03 | −0.06 | −0.06 | −0.01 |
| | | Difference of humidity from the current status (%) | 0.16 | 0.26 | 0.22 | 0.1 |
| | Solar gains (Wh/m²) | | | 0 | 0 | 0 |
| | Percentage difference from the current status | | | 0 | 0 | 0 |
| | Sensible heating (Wh/m²) | | | −7304.22 | −7463.33 | −7706.14 |
| | Percentage difference from the current status (%) | | | −19.49 | −21.33 | −24.10 |
| | Sensible cooling (Wh/m²) | | | 3509.26 | 3484.5 | 3669.99 |
| | Percentage difference from the current status (%) | | | −21.28 | −20.87 | −17.71 |

- **EHT-M1. Internal thermal insulation addition (ceiling)**

  See Table A.16.

**Table A.16** Results of assessment of EHT-M1 strategy and percentage difference from the current status

| EHT-M1 | | | Zone A | Zone B | Zone C | Zone D |
|--------|---|---|--------|--------|--------|--------|
| | | | N–O | N–E | S–E | S–O |
| P12 | Comfort | Difference of operative temperature from the current status (°C) | 0.11 | 0.08 | 0.09 | 0.13 |
| | | Difference of humidity from the current status (%) | −0.17 | −0.07 | −0.13 | −0.22 |
| | Solar gains (Wh/m²) | | | 0 | 0 | 0 |
| | Percentage difference from the current status | | | 0 | 0 | 0 |
| | Sensible heating (Wh/m²) | | | −8872.05 | −8921.75 | −8829.91 |
| | Percentage difference from the current status (%) | | | −23.68 | −25.50 | −27.61 |
| | Sensible cooling (Wh/m²) | | | 2103.39 | 2084.28 | 1774.5 |
| | Percentage difference from the current status (%) | | | −12.76 | −12.48 | −8.56 |

- **EVF-M1. External thermal insulation addition (façade)**

See Table A.17.

**Table A.17** Results of assessment of EVF-M1 strategy and percentage difference from the current status

| EVF-M1 | | | Zone A | Zone B | Zone C | Zone D |
|---|---|---|---|---|---|---|
| | | | N–O | N–E | S–E | S–O |
| P01 | Comfort | Difference of operative temperature from the current status (°C) | 0.65 | 0.62 | 0.59 | 0.61 |
| | | Difference of humidity from the current status (%) | −1.51 | −1.45 | −1.4 | −1.43 |
| | Solar gains (Wh/m²) | | | 0 | 0 | 0 |
| | Percentage difference from the current status | | | 0 | 0 | 0 |
| | Sensible heating (Wh/m²) | | | −11,663.2 | −8543.78 | −7566.85 |
| | Percentage difference from the current status (%) | | | −49.27 | −51.06 | −52.90 |
| | Sensible cooling (Wh/m²) | | | 451.38 | 1099.57 | 1565.61 |
| | Percentage difference from the current status (%) | | | −3.53 | −8.10 | −9.13 |
| P02–P11 | Comfort | Difference of operative temperature from the current status (°C) | 0.66 | 0.63 | 0.61 | 0.64 |
| | | Difference of humidity from the current status (%) | −1.53 | −1.47 | −1.45 | −1.5 |
| | Solar gains (Wh/m²) | | | 0 | 0 | 0 |
| | Percentage difference from the current status | | | 0 | 0 | 0 |
| | Sensible heating (Wh/m²) | | | −12,583.26 | −9736.13 | −8704.66 |
| | Percentage difference from the current status (%) | | | −48.92 | −51.18 | −53.29 |
| | Sensible cooling (Wh/m²) | | | 85.43 | 633.3 | 1061.66 |
| | Percentage difference from the current status (%) | | | −0.73 | −5.12 | −6.72 |
| P12 | Comfort | Difference of operative temperature from the current status (°C) | 0.33 | 0.36 | 0.34 | 0.31 |
| | | Difference of humidity from the current status (%) | −0.7 | −0.74 | −0.71 | −0.68 |
| | Solar gains (Wh/m²) | | | 0 | 0 | 0 |
| | Percentage difference from the current status | | | 0 | 0 | 0 |
| | Sensible heating (Wh/m²) | | | −10,207.7 | −9496.88 | −8994.75 |
| | Percentage difference from the current status (%) | | | −27.24 | −27.14 | −28.13 |
| | Sensible cooling (Wh/m²) | | | 141.5 | 300.09 | 18,822.43 |
| | Percentage difference from the current status (%) | | | −0.86 | −1.80 | −90.82 |

- **EVF-M2. Inside thermal insulation addition (façade air chamber)**

  See Table A.18.

**Table A.18** Results of assessment of EVF-M2 strategy and percentage difference from the current status

| EVF-M2 | | | Zone A | Zone B | Zone C | Zone D |
|---|---|---|---|---|---|---|
| | | | N–O | N–E | S–E | S–O |
| P01 | Comfort | Difference of operative temperature from the current status (°C) | 0.34 | 0.32 | 0.3 | 0.32 |
| | | Difference of humidity from the current status (%) | −0.78 | −0.72 | −0.7 | −0.75 |
| | Solar gains (Wh/m$^2$) | | | 0 | 0 | 0 |
| | Percentage difference from the current status | | | 0 | 0 | 0 |
| | Sensible heating (Wh/m$^2$) | | | −6879.23 | −5045.53 | −4600.47 |
| | Percentage difference from the current status (%) | | | −29.06 | −30.15 | −32.16 |
| | Sensible cooling (Wh/m$^2$) | | | 222.52 | 581.75 | 859.3 |
| | Percentage difference from the current status (%) | | | −1.74 | −4.29 | −5.01 |
| P02– P11 | Comfort | Difference of operative temperature from the current status (°C) | 0.34 | 0.32 | 0.31 | 0.34 |
| | | Difference of humidity from the current status (%) | −0.78 | −0.73 | −0.72 | −0.78 |
| | Solar gains (Wh/m$^2$) | | | 0 | 0 | 0 |
| | Percentage difference from the current status | | | 0 | 0 | 0 |
| | Sensible heating (Wh/m$^2$) | | | −7349.26 | −5699.95 | −5240.69 |
| | Percentage difference from the current status (%) | | | −28.57 | −29.96 | −32.08 |
| | Sensible cooling (Wh/m$^2$) | | | 21.53 | 334.3 | 577.15 |
| | Percentage difference from the current status (%) | | | −0.18 | −2.71 | −3.66 |
| P12 | Comfort | Difference of operative temperature from the current status (°C) | 0.2 | 0.2 | 0.19 | 0.18 |
| | | Difference of humidity from the current status (%) | −0.41 | −0.41 | −0.39 | −0.39 |
| | Solar gains (Wh/m$^2$) | | | 0 | 0 | 0 |
| | Percentage difference from the current status | | | 0 | 0 | 0 |
| | Sensible heating (Wh/m$^2$) | | | −5874.57 | −5427.58 | −5275.95 |
| | Percentage difference from the current status (%) | | | −15.68 | −15.51 | −16.50 |
| | Sensible cooling (Wh/m$^2$) | | | 53.95 | 137.71 | 445.06 |
| | Percentage difference from the current status (%) | | | −0.33 | −0.82 | −2.15 |

- **EVF-M3. Internal thermal insulation addition (façade)**

See Table A.19.

**Table A.19** Results of assessment of EVF-M3 strategy and percentage difference from the current status

| EVF-M3 | | | Zone A | Zone B | Zone C | Zone D |
|---|---|---|---|---|---|---|
| | | | N–O | N–E | S–E | S–O |
| P01 | Comfort | Difference of operative temperature from the current status (°C) | 0.34 | 0.33 | 0.31 | 0.32 |
| | | Difference of humidity from the current status (%) | −0.76 | −0.72 | −0.7 | −0.73 |
| | Solar gains (Wh/m²) | | | 0 | 0 | 0 |
| | Percentage difference from the current status (%) | | | 0.00 | 0.00 | 0.00 |
| | Sensible heating (Wh/m²) | | | −7098.69 | −5206 | −4694.15 |
| | Percentage difference from the current status (%) | | | −29.99 | −31.11 | −32.82 |
| | Sensible cooling (Wh/m²) | | | 49.1 | 394.84 | 485.65 |
| | Percentage difference from the current status (%) | | | −0.38 | −2.91 | −2.83 |
| P02–P11 | Comfort | Difference of operative temperature from the current status (°C) | 0.34 | 0.33 | 0.32 | 0.34 |
| | | Difference of humidity from the current status (%) | −0.77 | −0.73 | −0.72 | −0.76 |
| | Solar gains (Wh/m²) | | | 0 | 0 | 0 |
| | Percentage difference from the current status (%) | | | 0.00 | 0.00 | 0.00 |
| | Sensible heating (Wh/m²) | | | −7578.04 | −5875.23 | −5355.41 |
| | Percentage difference from the current status (%) | | | −29.46 | −30.88 | −32.79 |
| | Sensible cooling (Wh/m²) | | | −159.12 | 141.56 | 206.48 |
| | Percentage difference from the current status (%) | | | 1.36 | −1.15 | −1.31 |
| P12 | Comfort | Difference of operative temperature from the current status (°C) | 0.19 | 0.2 | 0.19 | 0.18 |
| | | Difference of humidity from the current status (%) | −0.38 | −0.39 | −0.38 | −0.36 |
| | Solar gains (Wh/m²) | | | 0 | 0 | 0 |
| | Percentage difference from the current status (%) | | | 0.00 | 0.00 | 0.00 |
| | Sensible heating (Wh/m²) | | | −6061.89 | −5582.13 | −5332.16 |
| | Percentage difference from the current status (%) | | | −16.18 | −15.95 | −16.67 |
| | Sensible cooling (Wh/m²) | | | −252.71 | −141.88 | −132.7 |
| | Percentage difference from the current status (%) | | | 1.53 | 0.85 | 0.64 |

- **EVPT-M1. External thermal insulation addition (thermal bridges)**

  See Table A.20.

**Table A.20** Results of assessment of EVPT-M1 strategy and percentage difference from the current status

| EVPT-M1 | | | Zone A | Zone B | Zone C | Zone D |
|---|---|---|---|---|---|---|
| | | | N–O | N–E | S–E | S–O |
| P01 | Comfort | Difference of operative temperature from the current status (°C) | 0.1 | 0.09 | 0.1 | 0.1 |
| | | Difference of humidity from the current status (%) | −0.23 | −0.22 | −0.23 | −0.24 |
| | Solar gains (Wh/m²) | | | 0 | 0 | 0 |
| | Percentage difference from the current status (%) | | | 0.00 | 0.00 | 0.00 |
| | Sensible heating (Wh/m²) | | | −2330.01 | −1845.22 | −1637.86 |
| | Percentage difference from the current status (%) | | | −9.84 | −11.03 | −11.45 |
| | Sensible cooling (Wh/m²) | | | −23.46 | 61.97 | 141.4 |
| | Percentage difference from the current status (%) | | | 0.18 | −0.46 | −0.83 |
| P02–P11 | Comfort | Difference of operative temperature from the current status (°C) | 0.1 | 0.09 | 0.1 | 0.1 |
| | | Difference of humidity from the current status (%) | −0.23 | −0.22 | −0.23 | −0.24 |
| | Solar gains (Wh/m²) | | | 0 | 0 | 0 |
| | Percentage difference from the current status (%) | | | 0.00 | 0.00 | 0.00 |
| | Sensible heating (Wh/m²) | | | −2508.08 | −2057.27 | −1862.63 |
| | Percentage difference from the current status (%) | | | −9.75 | −10.81 | −11.40 |
| | Sensible cooling (Wh/m²) | | | −45.12 | 20.79 | 87.78 |
| | Percentage difference from the current status (%) | | | 0.39 | −0.17 | −0.56 |
| P12 | Comfort | Difference of operative temperature from the current status (°C) | 0.06 | 0.07 | 0.06 | 0.06 |
| | | Difference of humidity from the current status (%) | −0.13 | −0.13 | −0.13 | −0.13 |
| | Solar gains (Wh/m²) | | | 0 | 0 | 0 |
| | Percentage difference from the current status (%) | | | 0.00 | 0.00 | 0.00 |
| | Sensible heating (Wh/m²) | | | −2082.37 | −1991.35 | −1869.63 |
| | Percentage difference from the current status (%) | | | −5.56 | −5.69 | −5.85 |
| | Sensible cooling (Wh/m²) | | | −42.45 | −18.85 | 65.59 |
| | Percentage difference from the current status (%) | | | 0.26 | 0.11 | −0.32 |

- **Improvement of insulation strategies comparison**

  See Table A.21.

**Table A.21** Results of improvement of insulation strategies assessment and comparison

| | | | EA | [M] Improvement of insulation | | | | | |
| | | | | EHCP | EHT | EVF | EVH | | EVPT |
| | | | | M1 | M1 | M1 | M2 | M3 | M1 |
| Comfort | Temperature (°C) | Air temperature | 22.04 | 22.04 | 22.04 | 22.49 | 22.26 | 22.26 | 22.11 |
| | | Radiant temperature | 22.31 | 22.31 | 22.32 | 23.06 | 22.71 | 22.72 | 22.43 |
| | | Operative temperature | 22.17 | 22.17 | 22.18 | 22.77 | 22.49 | 22.49 | 22.17 |
| | Relative humidity (%) | | 53.72 | 53.73 | 53.7 | 52.31 | 53 | 53.01 | 53.5 |
| Indoor gains (Wh/m$^2$) | Solar gains (windows) | | 37,147 | 37,147 | 37,147 | 37,147 | 37,147 | 37,147 | 37,147 |
| | Sensible heating of zone | | 21,043 | 20,441 | 20,291 | 11,026 | 15,122 | 14,969 | 18,989 |
| | Sensible cooling of zone | | −13,640 | −13,349 | −13,470 | −13,150 | −13,384 | −13,654 | −13,630 |
| Emissions (kg $CO_2$ equiv) | $CO_2$ production | | 35,574 | 35,089 | 35,027 | 28,834 | 31,600 | 31,583 | 34,219 |
| | Heating percentage of savings from the current status (%) | | | 1.4 | 1.5 | 18.9 | 11.2 | 11.2 | 3.8 |
| Consumptions (Wh/m$^2$) | Heating consumptions savings from the current status | | 22,873 | 22,219 | 22,055 | 11,985 | 16,437 | 16,271 | 20,640 |
| | Cooling percentage of savings from the current status (%) | | | 2.9 | 3.6 | 47.6 | 28.1 | 28.9 | 9.8 |
| | Cooling consumptions savings from the current status | | 6922 | 6777 | 6837 | 6688 | 6800 | 6938 | 6919 |
| | Total HVAC percentage of savings from the current status (%) | | | 2.1 | 1.2 | 3.4 | 1.8 | −0.2 | 0.0 |

# Appendix B

## B.1  Selected Social Indicators

See Table B.1.

**Table B.1** List of indicators, selection and social assessment of B1 information module results in its current status

| Module B1 | | Use | A | b | c | d |
|---|---|---|---|---|---|---|
| **e** | **f** | | | | | |
| Scenario description | | | Current status of building | | | |
| Aspect/indicator number | Aspect according to section | Indicator | | | | |
| 1 | Accessibility | 7.2.2.1 Approach to the building | | | | |
| | | 7.2.2.2 Access and movements inside the building | | | | |
| 2.9 | 7.2.2 | Dimensions and location of elevators | Y | Y — Elevator accessible 1.00 × 1.25 m | N | Substitution elevators study | CTE DB-SUA and Directive 293/2009 |
| 2 | Adaptability | 7.3 Adaptability | | | | |
| 1.1 | 7.3 | Ability of the building to accommodate individual user requirements | Y | N | N | New spatial redistributions possibilities | Law 38/1999 |
| 1.2 | 7.3 | Ability of the building to accommodate changes in user requirements | Y | N | N | New spatial redistributions possibilities | Law 38/1999 |
| 1.3 | 7.3 | Ability of the building to accommodate technical changes | Y | N | N | Upgrading obsolete facilities | Law 38/1999 |
| 1.4 | 7.3 | Ability of the building to accommodate the change of use | Y | N | N | List compatible activities | Law 38/1999 |
| 3 | Comfort and health | 7.4.2 Thermal characteristics | | | | |
| | | 7.4.2.1 Thermal characteristics according to structure of building | | | | |
| 1.1 | 7.4.2.1 | Operation temperature (°C or K) (radiant temperature of surfaces, air temperature and distribution) | Y | Y — T.° op. winter: 21–23 °C; T.° op. summer: 23–25 °C | N | Thermal renovation strategies and facilities improvements | RITE (IT 1.1.4.1.2.) |
| 1.2 | 7.4.2.1 | Humidity (% or g/kg) | Y | Y — H.R. winter: 40–50%; H.R. summer: 45–60% | N | Thermal renovation strategies and facilities improvements | RITE (IT 1.1.4.1.2.) |
| 1.3 | 7.4.2.1 | Air speed (m/s) and distribution | Y | Y — Air speed: 4 m/s; Summer ventilation: 4 r/h; Summer ventilation: 0.72 r/h | N | Air renovation system of building | Appendix C CTE DB-HE, CTE DB-HS3 and RITE |
| 1.4 | 7.4.2.1 | Type of activities in space | Y | Y — Sedentary metabolic activity 1.2 met | Y | Consider to calculations | RITE (IT 1.1.4.1.2.) |
| 1.5 | 7.4.2.1 | Type of users (clothing) | Y | Y — GV. summer: 0.5 clo; GV. winter: 0.5 clo | Y | – | RITE (IT 1.1.4.1.2.) |

(continued)

**Table B.1** (continued)

| Module B1 | | | Use | | | |
|---|---|---|---|---|---|---|
| e | f | | A | b | c | d |
| Scenario description | | | | | | |
| Aspect/indicator number | Aspect according to section | Current status of building — Indicator | | | | |
| | | 7.4.2.2 User and control system aspects of thermal characteristics | | | | |
| 2.2 | 7.4.2.2 | The operating temperature in individual rooms can be controlled (if yes: manually or automatically) [Yes/No] | Y | N | | |
| | | | Y | | Manual control: use of awnings, blinds, shutters, carpentry and operation of equipment | – |
| 2.5 | 7.4.2.2 | Humidity in individual rooms can be controlled (if yes: manually or automatically) [Yes/No] | Y | N | N | |
| | | | Y | | Manual control: use of awnings, blinds, shutters, carpentry and operation of equipment | – |
| 2.7 | 7.4.2.2 | Air velocity and air distribution in individual spaces can be controlled (if yes: manually or automatically) [Yes/No] | Y | N | | |
| | | | Y | | Manual | – |
| | | 7.4.3 Indoor air quality characteristics | | | | |
| | | 7.4.3.2 Indoor air quality aspects relating to the user and the control system | | | | |
| 4.2 | 7.4.3.2 | Is there user ventilation control by automatic and/or manual control? [Yes/No] | Y | N | N | |
| | | | Y | | Manual control: use of awnings, blinds, shutters, carpentry and operation of equipment | – |
| | | 7.4.4 Acoustic characteristics | | | | |
| 5.5 | 7.4.4 | Sound insulation of existing buildings | | | | |
| | | Consider the opinion of neighbours on this point and incorporate it into the intervention criteria if relevant | Y | Y | | |
| | | | N | | Conduct test and take corrective action | CTE DB-HR (only in integral refurbishment) and Standard EN ISO 16283-1 |
| | | 7.4.5 Visual comfort characteristics | | | | |
| | | 7.4.5.1 Aspects of the visual comfort characteristic related to the building structure | | | | |
| | | Contribution of natural light | | | | |
| 6.2 | 7.4.5.1 | Specific calculations for each room and activity to be studied | Y | Y | | |
| | | | N | | Lighting study for potential change of use (e.g. office) | Technical guide for the use of daylight in the lighting of buildings (IDAE) |
| 6.3 | 7.4.5.1 | Is there a visual connection to the outside world [Yes/No]? (1) What is the height of the windowsill (in metres)? (2) Is there a view of the different layers: sky, city/landscape and/or terrain? | Y | Y | | |
| | | Y (1) 1.10 m (2) Y, with the best views from the fifth floor upwards | N | | Transforming openings and enhancing the visual connection with the outside world | – |
| | | 7.4.5.2 Visual comfort aspects relating to the user and the control system | | | | |
| 7.2 | 7.4.5.2 | Can the user control the amount of daylight in individual spaces? [Yes/No] | Y | N | N | |
| | | | Y | | By means of curtains and blinds | – |

(continued)

**Table B.1** (continued)

| Module B1 | | Use | A | b | c | d |
|---|---|---|---|---|---|---|
| e | | | | | | |
| Scenario description | | | | | | |
| Aspect/indicator number | Aspect according to section | | Indicator | | | |
| | | | **Current status of building** | | | |
| | | | *7.4.6 Spatial characteristics* | | | |
| 8.2 | 7.4.6 | Y Y | Floor to ceiling height (m): 2.50 m | Y | Height reduction in centralised air-conditioning distributors and bathrooms | Art.12.2.29 of PGOU Malaga and CTE DB SUA/2 |
| 8.6 | 7.4.6 | Y Y | Outdoor space (type, e.g. balconies, terrace or garden and area ($m^2$): Maximum overhang = 1.50 m | Y | Increase the depth of existing terraces or create new ones | Art. 12.6.3 of PGOU Malaga CTE DB SE |
| | | | *7.5 Impacts on the neighbourhood* | | | |
| 11.3 | 7.5.3 | Y N | Water (e.g. drops from air-conditioning, water from gutters and downpipes): – | N | Re-routing to sewage disposal network | – |
| | | | *7.5.4 Glare/overshadowing* | | | |
| 12.1 | 7.5.4 | N N | Night glare: (1) the protection and illuminance (lux) of the test object at night and whether it is continuous or intermittent; (2) the presence of light (e.g. flickering, flashing, blinking, coloured, etc.) causing irritation, loss of concentration, etc. (1) Protection is continuous by means of manual shutter systems, and (2) lighting is continuous and uniform | N | Reduce levels of light pollution in the environment. There are problems with natural ventilation on summer nights | – |
| 12.2 | 7.5.4 | N | Day glare: (1) glare emitted by the surface of a building, e.g. caused by exterior materials with high reflectivity (1) Current materials do not cause glare to surrounding buildings | N | Interventions on the envelope do not affect this indicator | – |
| 12.3 | 7.5.4 | Y N | Overshadowing: (1) overshadowing with detrimental effects on the neighbourhood (area and hours of overshadowing on neighbours) (1) No excessive overshadowing due to surrounding buildings is detected | N | Avoid overshadowing with terrace operations | – |
| | | | *7.6 Maintenance and maintainability* | | | |
| 14.1 | 7.6 | Y Y | Frequency and duration of routine maintenance (including cleaning), repairs, replacements and/or rehabilitation. There is no requirement for a building book that identifies maintenance and control of the building | N | Establish a building book indicating the maintenance of the existing building and regeneration operations | Resolution of 11 June 2013, of the directorate general of Registries and Notaries General |
| 14.2 | 7.6 | Y Y | Impacts on users' health and comfort during maintenance work. Health and Safety Management for Construction | N | Health and Safety Management for Renovation of Building | Royal Decree 1627/1997 |
| 14.3 | 7.6 | Y Y | User safety during maintenance work. Health and Safety Management for Construction | N | Health and Safety Management for Renovation of Building | Royal Decree 1627/1997 |
| 14.4 | 7.6 | Y Y | Ability to use the building (usability) while maintenance tasks are being carried out, e.g. as a ratio of the expected duration of maintenance and cleaning that cause disruption on days of normal use. Health and Safety Management for Construction | N | Health and Safety Management for Renovation of Building | Royal Decree 1627/1997 |
| 4 | Security | | 7.7.2 Resilience to the consequences of climate change 7.7.2.2 Rain resistance | | | |

(continued)

**Table B.1** (continued)

| Module B1 | | | Use | | b | c | d |
|---|---|---|---|---|---|---|---|
| e | | | A | | | | |
| Scenario description | | | | | | | |
| Aspect/indicator number | Aspect according to section | Indicator | Current status of building | | | | |
| | f | | | | | | |
| **Resistance to rain and heavy rains** | | | | | | | |
| 15.1 | 7.7.2.2 | EN 12865, EN 1027 and EN 12208 | Y | Y | N | Control by means of specific tests the installed constructive systems | Law 38/1999 |
| **Water evacuation capacity** | | | | | | | |
| 15.2 | 7.7.2.2 | Provide a rainwater drainage system | Y | Y | N | Provide a network for the reuse of rainwater collected by a rainwater harvesting system | CTE DB HS 5 |
| **7.7.2.3 Wind resistance** | | | | | | | |
| **Increased structural strength** | | | | | | | |
| 16.1 | 7.7.2.3 | EN 12865, EN 1027 and EN 12208 | Y | N | N | Structure suitability study | – |
| **Measures to prevent the detachment of the façade or elements of the façade** | | | | | | | |
| 16.2 | 7.7.2.3 | | Y | N | N | Counting and evaluation of installed equipment technical sheet | DiT of products used |
| **Measures to improve the airtightness of building envelopes against the wind** | | | | | | | |
| 16.3 | 7.7.2.3 | | Y | Y | N | Counting and evaluation of installed equipment technical sheet | DiT of products used |
| **7.7.2.5 Flood resistance** | | | | | | | |
| **Water tightness of building envelopes and basements** | | | | | | | |
| 18.3 | 7.7.2.5 | | Y | Y | N | Counting and evaluation of installed equipment technical sheet | DiT of products used |
| **7.7.2.6 Solar radiation resistance** | | | | | | | |
| **Solar control measures, such as shading (e.g. blinds, cornices, eaves, screens) and/or types of window glazing** | | | | | | | |
| 19.1 | 7.7.2.6 | UNE-EN 13561:2015 | Y | N | N | Require compliance of products with standards and take measures for control according to previous solar survey | – |
| **Ultraviolet filters** | | | | | | | |
| 19.2 | 7.7.2.6 | UNE-EN 410:2011 | Y | N | N | Require compliance of products with standards and take measures for control according to previous solar survey | – |
| **Air conditioning, ventilation systems** | | | | | | | |
| 19.5 | 7.7.2.6 | | – | – | – | – | RITE |
| **Thermal inertia** | | | | | | | |
| 19.6 | 7.7.2.6 | | Y | N | N | Study of transformable blind panels | CTE |

# B.2 Allocation of Influence

See Table B.2.

**Table B.2** Allocation of influences in relation to selected indicator according to UNE-EN 16309 +A1 methodology

| Allocation of influence | | Aspect or impact indicator for the building information module B1 | Provision, measure or activity | Influence on the module | | | | | |
|---|---|---|---|---|---|---|---|---|---|
| | | | | Y | Y | Y | Y | Y | NAM |
| Aspect/indicator number | Aspect according to section | Indicator | Provision, measure or specific activity | B2 | B3 | B4 | B5 | B6 | B7 |
| 1 | Accessibility | 7.2.2.1 Approach to the building | | | | | | | |
| | | 7.2.2.2 Access and movement in the building | | | | | | | |
| 2.9 | 7.2.2 | The location, dimensions and ease lifts operation | Replacement of lifts | X | X | X | X | X | – |
| 2 | Adaptability | 7.3 Adaptability | | | | | | | |
| 1.1 | 7.3 | The ability of the building to accommodate the individual requirements of users | Enabling new redeployments | NAI | NAI | NAI | X | NAI | – |
| 1.2 | 7.3 | The ability of the building to accommodate changes in users' requirements | Enabling new redeployments | NAI | NAI | NAI | X | NAI | – |
| 1.3 | 7.3 | The ability of the building to accommodate technical changes | Updating obsolete installations | NAI | X | X | X | X | – |
| 1.4 | 7.3 | The ability of the building to accommodate change of use | Inventory compatible activities | NAI | NAI | NAI | X | X | – |
| 3 | Health and comfort | 7.4.2 Thermal characteristics | | | | | | | |
| | | 7.4.2.1 Aspects of thermal characteristics relating to the building fabric | | | | | | | |
| 1.1 | 7.4.2.1 | Operating temperature (°C or K) (radiant temperature of surfaces, air temperature and its distribution) | Adopt measures to transform the building thermal envelope | NAI | NAI | NAI | X | X | – |
| 1.2 | 7.4.2.1 | Humidity (% or g/kg) | | NAI | NAI | NAI | X | X | – |
| 1.3 | 7.4.2.1 | Air velocity (m/s) and distribution | Renovate the building's natural ventilation system | X | X | X | X | X | – |

(continued)

**Table B.2** (continued)

| Allocation of influence | | Aspect or impact indicator for the building information module B1 | Provision, measure or activity | Influence on the module | | | | | |
| --- | --- | --- | --- | --- | --- | --- | --- | --- | --- |
| | | | | Y | Y | Y | Y | Y | NAM |
| Aspect/indicator number | Aspect according to section | Indicator | Provision, measure or specific activity | B2 | B3 | B4 | B5 | B6 | B7 |
| 1.4 | 7.4.2.1 | Type of activities in the space ePara> | – | NAI | NAI | NAI | NAI | NAI | – |
| 1.5 | 7.4.2.1 | Type of users (clothing) | – | NAI | NAI | NAI | NAI | NAI | – |
| 7.4.2.2 User and control system aspects of thermal characteristics | | | | | | | | | |
| 2.2 | 7.4.2.2 | Operating temperature in individual spaces can be controlled (if yes: manually or automatically) [Yes/No] | Manual control measures: use of awnings, blinds, shutters and windows | X | X | X | X | NAI | – |
| 2.5 | 7.4.2.2 | Humidity in individual spaces can be controlled (if yes: manually or automatically) [Yes/No] | Manual control measures: use of awnings, blinds, shutters and windows | NAI | NAI | NAI | X | NAI | – |
| 2.7 | 7.4.2.2 | Air velocity and air distribution in individual spaces can be controlled (if yes: manually or automatically) [Yes/No] | Manual control measures: use of awnings, blinds, shutters and windows | X | X | X | X | NAI | – |
| 7.4.3 Indoor air quality characteristics | | | | | | | | | |
| 7.4.3.2 Indoor air quality aspects relating to the user and the control system | | | | | | | | | |
| 4.2 | 7.4.3.2 | Is user ventilation controlled by automatic and/or manual control? [Yes/No] | Manual control measures: use of awnings, blinds, shutters and windows | NAI | NAI | X | X | X | – |
| 7.4.4 Acoustic characteristics | | | | | | | | | |
| 5.5 | 7.4.4 | Sound insulation of the existing buildings | Carry out test and take corrective measures | NR | NAI | NAI | X | NAI | – |
| 7.4.5 Visual comfort characteristics | | | | | | | | | |
| 7.4.5.1 Aspects of the visual comfort characteristic related to the building structure | | | | | | | | | |
| 6.2 | 7.4.5.1 | Contribution of natural light | Lighting study to support potential changes of use (e.g. offices) | NAI | NAI | NAI | X | X | – |
| 6.3 | 7.4.5.1 | Is there a visual connection to the outside world [Yes/No]? (1) What is the height of the windowsill (in metres)? (2) Is | Transform openings and enhance visual connection with the exterior | NAI | NAI | NAI | X | NAI | – |

(continued)

**Table B.2** (continued)

| Allocation of influence | | Aspect or impact indicator for the building information module B1 | Provision, measure or activity | Influence on the module | | | | | |
|---|---|---|---|---|---|---|---|---|---|
| | | | | Y | Y | Y | Y | Y | NAM |
| Aspect/indicator number | Aspect according to section | Indicator | Provision, measure or specific activity | B2 | B3 | B4 | B5 | B6 | B7 |
| | | there a view of the different layers: sky, city/landscape and/or terrain? | | | | | | | |
| 7.4.5.2 Visual comfort aspects concerning the user and the control system | | | | | | | | | |
| 7.2 | 7.4.5.2 | Can the user control the amount of daylighting in the individual spaces?—[Yes/No] | Manual control measures: use of awnings, blinds, shutters and windows | NAI | NAI | NAI | X | NAI | – |
| 7.4.6 Spatial characteristics | | | | | | | | | |
| 8.2 | 7.4.6 | Floor to ceiling height (m) | Reduction of the height of the hallways and bathrooms for the implementation of a centralised climate control system in the house | NAI | NAI | NAI | X | NAI | – |
| 8.6 | 7.4.6 | Outdoor space (type, e.g. balconies, terrace or garden and area) (m²) | Possibility of increasing the depth of existing terraces or creating new ones | NAI | NAI | NAI | X | NAI | – |
| 7.5 Impacts on the neighbourhood | | | | | | | | | |
| 11.3 | 7.5.3 | Water (e.g. air-conditioning drops, water from gutters and downpipes) | Redirect to water drainage network | NR | NAI | NAI | X | NAI | – |
| 7.5.4 Glare/overshadowing | | | | | | | | | |
| 12.1 | 7.5.4 | Night glare: (1) the protection and illuminance (lux) of the assessment object at night and whether it is continuous or intermittent; (2) the presence of light (e.g. flickering, flashing, blinking, coloured) that causes irritation, loss of concentration, etc. | Reduce levels of light pollution in the surroundings. There are problems with natural ventilation on summer nights | NAI | NAI | NAI | NAI | NAI | – |

(continued)

**Table B.2** (continued)

| Allocation of influence | | Aspect or impact indicator for the building information module B1 | Provision, measure or activity | Influence on the module | | | | | |
|---|---|---|---|---|---|---|---|---|---|
| | | | | Y | Y | Y | Y | Y | NAM |
| Aspect/indicator number | Aspect according to section | Indicator | Provision, measure or specific activity | B2 | B3 | B4 | B5 | B6 | B7 |
| 12.2 | 7.5.4 | Daytime glare: (1) glare emitted by the surface of a building, e.g. caused by exterior materials with high reflectivity | Ensure that interventions at the envelope level do not affect this indicator | NAI | NAI | NAI | NAI | NAI | – |
| 12.3 | 7.5.4 | Overshadowing: (1) overshadowing with detrimental effects on the neighbourhood (area and hours of overshadowing on neighbours) | Avoid overshadowing with operations on terraces | NAI | NAI | NAI | NAI | NAI | – |
| | | 7.6 Maintenance and maintainability | | | | | | | |
| 14.1 | 7.6 | Frequency and duration of routine maintenance (including cleaning), repairs, replacements and/or rehabilitation | Draw up a building book indicating routine maintenance of the consolidated work and regeneration operations | X | X | X | X | NAI | – |
| 14.2 | 7.6 | Impacts on health and comfort of users during maintenance work | Health and Safety Management for Construction perform | X | X | X | X | NAI | – |
| 14.3 | 7.6 | Safety of users during maintenance tasks | Health and Safety Management for Construction perform | X | X | X | X | NAI | – |
| 14.4 | 7.6 | Ability to use the building (usability) while maintenance tasks are carried out, e.g. as a ratio of the expected duration of maintenance and cleaning that cause disruption on days of normal use | Health and Safety Management for Construction perform during the building regeneration works | X | X | X | X | NAI | – |
| 4 | Security | 7.7.2 Resilience to the consequences of climate change | | | | | | | |
| | | 7.7.2.2 Resistance to rainfall | | | | | | | |
| 15.1 | 7.7.2.2 | Resistance to rainfall and heavy rainfall | Control by means of specific tests of the construction | NAI | NAI | NAI | X | NAI | – |

(continued)

**Table B.2** (continued)

| Allocation of influence | | Aspect or impact indicator for the building information module B1 | Provision, measure or activity | Influence on the module | | | | | |
|---|---|---|---|---|---|---|---|---|---|
| | | | | Y | Y | Y | Y | Y | NAM |
| Aspect/indicator number | Aspect according to section | Indicator | Provision, measure or specific activity | B2 | B3 | B4 | B5 | B6 | B7 |
| | | | systems installed | | | | | | |
| 15.2 | 7.7.2.2 | Water drainage capacity | Provide separate network for rainwater reuse | NAI | NAI | NAI | X | NAI | – |
| 7.7.2.3 Wind resistance | | | | | | | | | |
| 16.1 | 7.7.2.3 | Increased structural strength | Testing the structure | NAI | NAI | NAI | NAI | NAI | – |
| 16.2 | 7.7.2.3 | Measures to prevent detachment of the facade or facade elements | Count and evaluate the data sheets of the materials installed | NAI | NAI | NAI | X | NAI | – |
| 16.3 | 7.7.2.3 | Measures to improve the airtightness of building envelopes against wind | Count and evaluate the data sheets of the materials installed | NAI | NAI | NAI | X | NAI | – |
| 7.7.2.5 Resistance to flooding | | | | | | | | | |
| 18.3 | 7.7.2.5 | Watertightness of building envelopes and basements | Count and evaluate the data sheets of the materials installed | NAI | NAI | NAI | X | NAI | – |
| 7.7.2.6 Resistance to solar radiation | | | | | | | | | |
| 19.1 | 7.7.2.6 | Solar control measures, such as shading (e.g. blinds, cornices, eaves, screens) and/or types of window glazing | Require compliance with standards for installed products and adopt solar control measures based on a prior solar study | NAI | NAI | NAI | X | NAI | – |
| 19.2 | 7.7.2.6 | Ultraviolet filters | Require compliance with standards in relation to the light and solar characteristics of glazing | NR | NAI | NAI | X | NAI | – |
| 19.5 | 7.7.2.6 | Air conditioning, ventilation systems | – | X | X | X | X | X | – |
| 19.6 | 7.7.2.6 | Thermal inertia | Study the incorporation of blind panels in the envelope that could function as Trombe walls | NAI | NAI | NAI | X | X | – |

# Appendix C

## C.1 Environmental and Economic Impacts Calculations

See Tables C.1, C.2 and C.3.

**Table C.1** Determination of environmental and economic impacts of the different design options (DOCs) for strategy EVF-B4 + M1

| EVF-B4 + M1 | B4 | Transformation into a ventilated façade | | | | | | | |
| | M1 | External thermal insulation addition (façade) | | | | | | | |
| Measurement (m²) and strategy code | | Design options (DO) | | | Environmental indicators | | | | |
| | | Options (component level) | | Options (subcomponent level) | Construction (A1–A5) | | | | Transport (C2) |
| | | | | | ADP_fossil fuels (MJ) | | GWP (kg CO₂ equiv) | | |
| | | | | | By m² | MJ | By m² | kg CO₂ | kg CO₂ |
| EV | 3037.3 | B4 | OP1 | Cladding for ventilated façade formation with HPL high-pressure decorative laminate panel | Decorative high-pressure laminate panel HPL | 172.48 | 523,873.50 | 10.23 | 31,071.58 | |
| | | | | | Aluminium profiles for ventilated façade formation | 194.93 | 592,060.89 | 11.10 | 33,714.03 | |
| | | | | | | **367.41** | **1,115,934.39** | **21.33** | **64,785.61** | **1612.00** |
| | | | OP2 | Cladding for the formation of a ventilated façade with single-sided polymer concrete elements | Polymer concrete part for single-sided façade | 70.29 | 213,491.82 | 6.68 | 20,289.16 | |
| | | | | | Aluminium supporting structure for ventilated façade formation with concrete parts | 250.35 | 760,388.06 | 14.36 | 43,615.63 | |
| | | | | | | **320.64** | **973,879.87** | **21.04** | **63,904.79** | **1612.00** |
| | | | OP3 | Cladding for the formation of a ventilated façade with ceramic tiles | Single-sided extruded porcelain stoneware tile façade tile | 61.91 | 188,039.24 | 4.70 | 14,275.31 | |
| | | | | | Aluminium supporting structure for ventilated façade with ceramic tiles | 692.16 | 2,102,297.57 | 39.41 | 119,699.99 | |
| | | | | | | **754.07** | **2,290,336.81** | **44.11** | **133,975.30** | **2184.70** |
| | | M1 | OP1 | | Rigid glass mineral wool (MW) board for insulation purposes | 142.07 | 431,509.21 | 4.09 | 12,422.56 | |
| | | | | | Nylon dowel and support for fixing insulating materials | 1.13 | 3432.15 | 0.17 | 516.34 | |
| | | | | | | **143.20** | **434,941.36** | **4.26** | **12,938.90** | **1612.00** |
| | | | OP2 | | Expanded polystyrene sheet (EPS) | 73.71 | 223,879.38 | 10.88 | 33,045.82 | |
| | | | | | Nylon dowel and support for fixing insulating materials | 3.01 | 9,142.27 | 0.44 | 1,336.41 | |
| | | | | | | **76.72** | **233,021.66** | **11.32** | **34,382.24** | **1612.00** |
| | | | OP3 | | Extruded polystyrene sheet (XPS) | 147.43 | 447,789.14 | 21.76 | 66,091.65 | |
| | | | | | Nylon plug and nylon support for fixing insulating materials | 3.01 | 9142.27 | 0.44 | 1336.41 | |
| | | | | | | **150.44** | **456,931.41** | **22.20** | **67,428.06** | **1612.00** |

(continued)

**Table C.1** (continued)

Transformation into a ventilated façade

External thermal insulation addition (façade)

Economic indicators

| Construction (A1–A5) | | Transport, CDW treatment and disposal (C2–C4) | | | | | | | | | | | Total, modules C2–C4 | Costs beyond the system boundary (module D$_{B2–B4}$) | |
|---|---|---|---|---|---|---|---|---|---|---|---|---|---|---|---|
| Non annual costs (€) | | Transport (C2) to landfill (€) | | CDW treatment cost (C3) | | Disposal tax (C4) | | | Estimation of expected amount of CDW | | | | | Decennial maintenance | |
| By m² | € | By m³ | € | By m³ | € | Type of CDW | By m³ | € | Tn/m² | Tn | m³/m² | m³ | € | By m² | €/year |
| | | 1.91 | 0.75 | | | Unsorted mixture | 18.41 | 7.21 | 1.94E-04 | 0.59 | 1.29E-04 | 0.39 | | | |
| | | 2.06 | 0.33 | | | Plastics | 27.38 | 4.41 | 3.20E-05 | 0.10 | 5.30E-05 | 0.16 | | | |
| 128.1 | 389,108.50 | | **1.08** | 15.00 | **8.29** | | | **11.62** | | **0.69** | | **0.55** | **20.99** | 9.99 | 3034.26 |
| | | 3.64 | 6.12 | | | Concretes | 8.25 | 13.88 | 6.93E-04 | 2.10 | 5.54E-04 | 1.68 | | | |
| | | 11.13 | 14.81 | | | Metals | 15.66 | 20.83 | 3.82E-04 | 1.16 | 4.38E-04 | 1.33 | | | |
| 133.6 | 405,752.91 | | **20.93** | 15.00 | **45.20** | | | **34.72** | | **3.27** | | **3.01** | **100.84** | 8.80 | 2672.82 |
| | | 3.16 | 12.23 | | | Ceramic materials | 8.25 | 31.92 | 1.59E-03 | 4.84 | 1.27E-03 | 3.87 | | | |
| | | 11.13 | 13.12 | | | Metals | 15.66 | 18.45 | 3.35E-04 | 1.02 | 3.88E-04 | 1.18 | | | |
| 111.2 | 337,687.01 | | **25.34** | 15.00 | **75.72** | | | **50.38** | | **5.86** | | **5.05** | **151.44** | 25.6 | 7766.38 |
| | | 1.91 | 0.78 | | | Non-classified mixture | 18.41 | 7.55 | 8.10E-05 | 0.25 | 1.35E-04 | 0.41 | | | |
| | | 2.06 | 0.26 | | | Plastics | 27.38 | 3.49 | 2.50E-04 | 0.76 | 4.20E-05 | 0.13 | | | |
| 11.28 | 34,260.74 | | **1.05** | 0.00 | **0.00** | | | **11.04** | | **1.01** | | **0.54** | **12.09** | 0.23 | 69.86 |
| | | 1.91 | 0.44 | | | Non-classified mixture | 18.41 | 4.19 | 4.50E-05 | 0.14 | 7.50E-05 | 0.23 | | | |
| | | 2.06 | 0.26 | | | Plastics | 27.38 | 3.49 | 2.50E-05 | 0.08 | 4.20E-05 | 0.13 | | | |
| 8.82 | 26,788.99 | | **0.70** | 0.00 | **0.00** | | | **7.69** | | **0.21** | | **0.36** | **8.38** | 0.18 | 54.67 |

(continued)

# Table C.1 (continued)

**Transformation into a ventilated façade**

**External thermal insulation addition (façade)**

**Economic indicators**

| Construction (A1–A5) | | Transport, CDW treatment and disposal (C2–C4) | | | | | | | | | | | | Costs beyond the system boundary (module $D_{B2–B4}$) | |
|---|---|---|---|---|---|---|---|---|---|---|---|---|---|---|---|
| | | Transport (C2) to landfill (€) | | CDW treatment cost (C3) (€) | | Disposal tax (C4) | | | Estimation of expected amount of CDW | | | | Total, modules C2–C4 | Decennial maintenance | |
| Non annual costs (€) | | | | | | | | | | | | | | | |
| By m² | € | By m³ | € | By m³ | € | Type of CDW | By m³ | € | Tn/m² | Tn | m³/m² | m³ | € | By m² | €/year |
| | | | | | | Non-classified mixture | 18.41 | 7.16 | 7.70E–05 | 0.23 | 1.28E–04 | 0.39 | | | |
| | | 1.91 | 0.74 | | | Plastics | 27.38 | 3.49 | 2.50E–05 | 0.08 | 4.20E–05 | 0.13 | | | |
| | | 2.06 | 0.26 | | | | | | | | | | | | |
| 8.30 | 25,209.59 | 1.01 | | 0.00 | 0.00 | | | 10.65 | | 0.31 | | 0.52 | 11.66 | 0.17 | 51.63 |

*Source* Own elaboration based on the contextualization of the data from the different sources of information analysed (IQM)

**Table C.2** Determination of environmental and economic impacts of different design options for the EHCP-B2 + M1 strategy at roof level

| EHCP-B2 + M1 | | | | | | Environmental indicators | | | | |
|---|---|---|---|---|---|---|---|---|---|---|
| B2 | Transformation in a ventilated flat roof | | | | | Construction (A1–A5) | | | | Transport (C2) |
| M1 | External thermal insulation addition (roof) | | | | | ADP_fossil fuels (MJ) | | GWP (kg CO₂ equiv) | | |
| Measurement (m²) and strategy code | | | Design options (DO) | Options (component level) | Options (subcomponent level) | By m² | MJ | By m² | kg CO₂ | kg CO₂ |
| EH | 419.3 | B2 | OP1 | Energy rehabilitation with filtering slab for the formation of a walkable roof with ventilated chamber | Prefabricated lightened and filtering concrete parts | 2191.8 | 919,030.13 | 217.18 | 91,063.57 | |
| | | | | | Geotextile made of polypropylene felt | 9.12 | 3824.02 | 1.35 | 566.06 | |
| | | | | | | **2200.9** | **922,854.14** | **218.53** | **91,629.63** | **1612.00** |
| | | | OP2 | Inverted walkable roof with slopes of terrazzo paving on supports | Waterproofing membrane | 245.06 | 102,753.66 | 36.01 | 15098.99 | |
| | | | | | Geotextile formed by polypropylene felt | 9.12 | 3824.02 | 1.35 | 566.06 | |
| | | | | | Rigid glass mineral wool (MW) slab for insulation | 9.12 | 3824.02 | 1.35 | 566.06 | |
| | | | | | Terrazzo paving system with boulder aggregate on plots | 323.10 | 135,475.83 | 30.83 | 12,927.02 | |
| | | | | | | **332.22** | **245,877.52** | **32.18** | **29,158.12** | **1612.00** |
| | | M1 | OP1 | | Rigid glass mineral wool (MW) board for insulation purposes | 142.07 | 59,569.95 | 4.09 | 1714.94 | |
| | | | | | | **142.07** | **59,569.95** | **4.09** | **1714.94** | **1612.00** |
| | | | OP2 | | Expanded polystyrene sheet (EPS) | 73.71 | 30,906.60 | 10.88 | 4561.98 | |
| | | | | | | **73.71** | **30,906.60** | **10.88** | **4561.98** | **1612.00** |
| | | | OP3 | | Extruded polystyrene sheet (XPS) | 147.43 | 61,817.40 | 21.76 | 9123.97 | |
| | | | | | | **147.43** | **61,817.40** | **21.76** | **9123.97** | **1612.00** |

(continued)

# Table C.2 (continued)

**Transformation in a ventilated flat roof**

**External thermal insulation addition (roof)**

Economic indicators

| Construction (A1–A5) | | Transport, CDW treatment and disposal (C2–C4) | | | | | | | Estimation of expected amount of CDW | | | | Total, modules C2–C4 | Costs beyond the system boundary (module $D_{B2-B4}$) | |
|---|---|---|---|---|---|---|---|---|---|---|---|---|---|---|---|
| Non annual costs (€) | | Transport (C2) to landfill (€) | | CDW treatment cost (C3) | | Disposal tax (C4) | | | | | | | | Decennial maintenance | |
| By m² | € | By m³ | € | By m³ | € | Type of CDW | By m³ | € | $Tn/m^2$ | Tn | $m^3/m^2$ | $m^3$ | € | By m² | €/year |
| 46.11 | 19,333.92 | 3.64 | 3.46 | 15.00 | | Concretes | 8.25 | 7.83 | 3.40E-03 | 1.42 | 2.26E-03 | 0.95 | | 30.37 | |
| | | 1.28 | 0.46 | | | Paper and cardboard | 15.66 | 5.63 | 6.43E-04 | 0.27 | 8.57E-04 | 0.36 | | | |
| | | | **3.92** | | **19.63** | | | **13.46** | | **1.69** | | **1.31** | **37.00** | | **1273.41** |
| | | 1.91 | 2.43 | 15.00 | | Non-classified mixture | 18.41 | 23.41 | 3.59E-03 | 1.50 | 3.03E-03 | 1.27 | | | |
| | | 2.06 | 0.01 | | | Plastics | 27.38 | 0.17 | 9.00E-06 | 0.00 | 1.50E-05 | 0.01 | | | |
| | | 2.06 | 0.16 | | | Plastics | 27.38 | 2.10 | 1.10E-04 | 0.05 | 1.83E-04 | 0.08 | | | |
| 133.6 | 56,014.29 | 1.28 | 1.73 | 15.00 | | Paper and cardboard | 15.66 | 42.43 | 5.63E-04 | 0.24 | 9.05E-04 | 1.35 | | 4.00 | |
| | | | **4.32** | | **40.64** | | | **68.12** | | **1.79** | | **2.71** | **113.08** | | **167.72** |

**Transformation in a ventilated flat roof**

**External thermal insulation addition (roof)**

Economic indicators

| Construction (A1–A5) | | Transport, CDW treatment and disposal (C2–C4) | | | | | | | Estimation of expected amount of CDW | | | | Total, modules C2–C4 | Costs beyond the system boundary (module $D_{B2-B4}$) | |
|---|---|---|---|---|---|---|---|---|---|---|---|---|---|---|---|
| Non annual costs (€) | | Transport (C2) to landfill (€) | | CDW treatment cost (C3) | | Disposal tax (C4) | | | | | | | | Decennial maintenance | |
| By m² | € | By m³ | € | By m³ | € | Type of CDW | By m³ | € | $Tn/m^2$ | Tn | $m^3/m^2$ | $m^3$ | € | By m² | €/year |
| 11.28 | 4729.70 | 1.91 | | 15.00 | | Non-classified mixture | 18.41 | 7.55 | 8.10E-05 | 0.03 | 1.35E-04 | 0.41 | | 0.23 | |
| | | | **0.78** | | **6.15** | | | **7.55** | | **0.03** | | **0.41** | **14.48** | | **9.64** |
| 8.82 | 3698.23 | 1.91 | | 15.00 | | Non-classified mixture | 18.41 | 4.19 | 4.50E-05 | 0.02 | 7.50E-05 | 0.23 | | 0.23 | |
| | | | **0.44** | | **3.42** | | | **4.19** | | **0.02** | | **0.23** | **8.05** | | **9.64** |
| 8.30 | 3480.19 | 1.91 | | 15.00 | | Non-classified mixture | 18.41 | 7.16 | 7.70E-05 | 0.03 | 1.28E-04 | 0.39 | | 0.23 | |
| | | | **0.74** | | **5.83** | | | **7.16** | | **0.03** | | **0.39** | **13.73** | | **9.64** |

*Source* Own elaboration based on the contextualization of the data from the different sources of information analysed (IQM)

**Table C.3** Determination of environmental and economic impacts of the different design options for the EVH C3 strategy at the void level

| EVH C3 | C3 | Addition of new carpentry to the exterior of the opening | | Environmental indicators | | | | |
| --- | --- | --- | --- | --- | --- | --- | --- | --- |
| | | | | Construction (A1–A5) | | | | Transport (C2) |
| | | | | ADP_fossil fuels (MJ) | | GWP (kg CO₂ equiv) | | |
| Measurement [Number of windows/balconies (NW/NB), windows/balconies surface (SW/SB)] and strategy code | Design options (DO) | | | By m² | MJ | By m² | kg CO₂ | kg CO₂ |
| | Options (component level) | | Options (subcomponent level) | | | | | |
| NW 240.00 NB 48.00 SW 290.40 SB 172.80 | C3 | OP1 Addition of 110 × 110 cm window, with non-plasticized PVC window | Sealing of joinery joint with the building opening (ml) | 1.81 | 1911.36 | 0.27 | 285.12 | |
| | | | Non-plasticized PVC window (unit) | 2430 | 583,252.80 | 333.21 | 79,970.40 | |
| | | | Installation of steel frame (pcs.) | 59.67 | 14,320.80 | 8.26 | 1982.40 | |
| | | | 6/12/6 mm clear glass insulating pane (m²) | 427.68 | 124,198.27 | 26.18 | 7602.67 | |
| | | | | 2919 | **723,683.23** | 367.92 | **89,840.59** | |
| | | Addition of 180 × 200 cm balcony window, with non-plasticized PVC balcony window | Sealing the joint between the joinery and the building opening (ml) | 4.03 | 1547.52 | 0.59 | 226.56 | |
| | | | Non-plasticized PVC balcony window (unit) | 7642 | 366,811.68 | 1033 | 49,597.44 | |
| | | | Installation of steel frame (unit) | 116.71 | 5602.08 | 16.73 | 803.04 | |
| | | | 6/12/3 + 3 mm clear glass insulating pane (m²) | 2091 | 361,302.34 | 128.00 | 22,118.40 | |
| | | | | 9854 | **735,263.62** | 1179 | **72,745.44** | |
| | | | **Totals** | | **1,458,946.8** | | **162,586.03** | **1612.00** |
| | | OP2 Addition of 110 × 110 cm window, with white lacquered aluminium window with thermal break | Sealing of the joinery joint with the building opening (ml) | 1.81 | 1911.36 | 0.27 | 285.12 | |
| | | | White lacquered aluminium window with thermal break (unit) | 3320 | 796,708.80 | 471.21 | 113,090.40 | |
| | | | Installation of steel frame (pcs.) | 59.67 | 14,320.80 | 8.26 | 1982.40 | |
| | | | 6/12/6 mm clear glass insulating pane (m²) | 427.68 | 124,198.27 | 26.18 | 7602.67 | |
| | | | | 3809 | **937,139.23** | 505.92 | **122,960.59** | |
| | | Addition of 180 × 220 cm balcony window, with white lacquered aluminium balcony window with thermal break | Sealing the joint between the carpentry and the building opening (ml) | 4.03 | 1547.52 | 0.59 | 226.56 | |
| | | | White lacquered aluminium balcony window with thermal break (unit) | 10,981 | 527,082.72 | 1559 | 74,838.24 | |
| | | | Fitting of steel frame (unit) | 116.71 | 5602.08 | 16.73 | 803.04 | |

(continued)

**Table C.3** (continued)

| EVH C3 | C3 | Addition of new carpentry to the exterior of the opening | | | | | |
|---|---|---|---|---|---|---|---|
| Measurement [Number of windows/balconies (NW/NB), windows/balconies surface (SW/SB)] and strategy code | | Design options (DO) Options (component level) | Options (subcomponent level) | Environmental indicators | | | |
| | | | | Construction (A1–A5) | | | Transport (C2) |
| | | | | ADP_fossil fuels (MJ) | | GWP (kg CO₂ equiv) | |
| | | | | By m² | MJ | By m² | kg CO₂ | kg CO₂ |
| | | | 6/12/3 + 3 mm clear glass insulating glazing (m²) | 2091 | 361,302.34 | | 22,118.40 | |
| | | | **Totals** | 13,193 | 895,534.66 | 1704 | 97,986.24 | **1612.00** |
| | | | | | **1,832,673.9** | | **220,946.83** | |

Addition of new carpentry to the exterior of the opening

Economic indicators

| Non annual costs (A1–A5) | | Transport, CDW treatment and disposal (C2–C4) | | | | | | | Estimation of expected amount of CDW | | | | Total, modules C2–C4 | Costs beyond the system boundary (module D_B2–B4) | |
|---|---|---|---|---|---|---|---|---|---|---|---|---|---|---|---|
| Construction (A1–A5) | | Transport (C2) to landfill | | CDW treatment cost (C3) | | Disposal tax (C4) | | | | | | | | Decennial maintenance | |
| By m² | € | By m³ | € | By m³ | € | Type of CDW | By m³ | € | Tn/m² | Tn | m³/m² | m³ | € | By m² | €/year |
| | | 1.91 | 0.01 | | | Non-classified mixture | 18.41 | 0.06 | 1.90E−05 | 0.00 | 1.30E−05 | 3E−03 | | | |
| | | 2.06 | 0.02 | | | Plastics | 27.38 | 0.23 | 2.08E−04 | 0.05 | 3.47E−05 | 8E−03 | | | |
| 363.5 | 87,232.80 | | 0.02 | | | | | 0.29 | | 0.05 | | 0.01 | | 32.7 | 785.04 |
| | | 1.91 | 0.01 | | | Non-classified mixture | 18.41 | 0.10 | 3.40E−05 | 0.01 | 2.30E−05 | 6E−03 | | | |
| | | 2.06 | 0.23 | | | Plastics | 27.38 | 3.01 | 2.75E−04 | 0.07 | 4.58E−04 | 1E−01 | | | |
| 661.2 | 31,739.04 | | 0.24 | | | | | 3.11 | | 0.07 | | 0.12 | | 59.5 | 285.65 |
| | **118,971.84** | | **0.26** | | | | | **3.21** | | **0.14** | | **0.14** | **3.47** | | **1070.69** |
| | | 1.91 | 0.01 | | | Non-classified mixture | 18.41 | 0.06 | 1.90E−05 | 0.00 | 1.30E−05 | 3E−03 | | | |

(continued)

**Table C.3** (continued)

Addition of new carpentry to the exterior of the opening

Economic indicators

| Construction (A1–A5) Non annual costs (€) | | Transport, CDW treatment and disposal (C2–C4) | | | | | | | | | | | | Costs beyond the system boundary (module $D_{B2–B4}$) Decennial maintenance | |
| | | Transport (C2) to landfill (€) | | CDW treatment cost (C3) | | Disposal tax (C4) | | | Estimation of expected amount of CDW | | | | Total, modules C2–C4 | | |
| By m² | € | By m³ | € | By m³ | € | Type of CDW | By m³ | € | Tn/m² | Tn | m³/m² | m³ | € | By m² | €/year |
|---|---|---|---|---|---|---|---|---|---|---|---|---|---|---|---|
| | | 2.06 | 0.02 | | | Plastics | 27.38 | 0.24 | 2.17E–04 | 0.05 | 3.62E–05 | 9E–03 | | | |
| 465.6 | **111,746.40** | 2.06 | 0.02 | | | | | 0.30 | | 0.06 | | 0.01 | | 51.2 | **1229.28** |
| | | 1.91 | 0.01 | | | Non-classified mixture | 18.41 | 0.10 | 3.40E–05 | 0.01 | 2.30E–05 | 6E–03 | | | |
| | | 2.06 | 0.24 | | | Plastics | 27.38 | 3.15 | 2.88E–04 | 0.07 | 4.80E–04 | 1E–01 | | | |
| 698.8 | **33,543.84** | | 0.25 | | | | | 3.26 | | 0.08 | | 0.12 | | 76.9 | **368.98** |
| | **145,290.24** | | **0.27** | | | | | **3.36** | | **0.15** | | **0.14** | **3.63** | | **1598.26** |

*Source* Own elaboration based on the contextualization of the data from the different sources of information analysed (IQM)

# Appendix D

## D.1 TERP Plans

See Figs. D.1 and D.2.

P. Mercader-Moyano and M. Ramos Martín, *Sustainable Renovation of Buildings*,
SpringerBriefs in Geography, https://doi.org/10.1007/978-3-031-15143-9

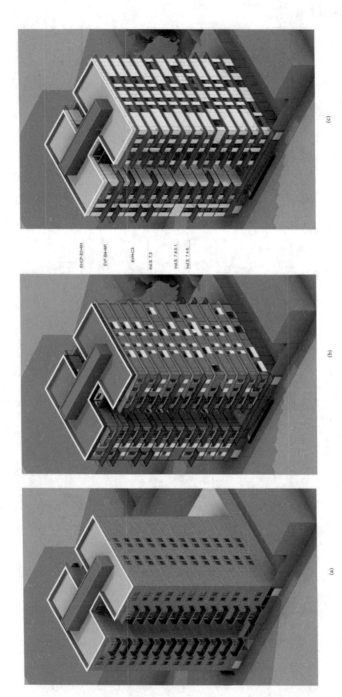

**Fig. D.1 a** Isometric perspective of the building model in its current state. **b** Isometric perspective of the building model according to a TERP, identifying regeneration actions. **c** Isometric perspective of the building model according to the TERP

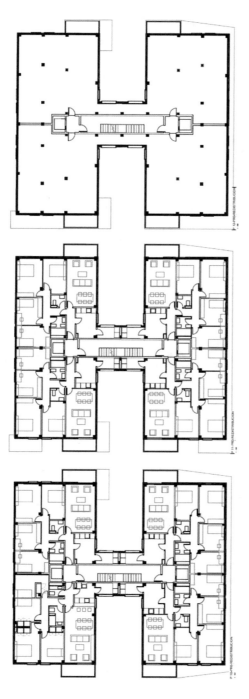

**Fig. D.2** Interventions and distribution options plans

# Index

**B**
Building Information Modelling (BIM)   1, 9, 15, 19, 21, 34

**C**
Case study   62, 63

**E**
Energy retrofitting   7, 15, 19

**F**
Future perspectives   19, 53, 58

**Q**
Qualification model   19

**Quantification model**   16, 19–21, 40, 52, 58, 59

**R**
Results   9, 15, 19, 24, 25, 28, 32, 35–37, 43–52, 54, 58, 59, 64–66, 68, 70, 72, 74, 76, 78, 79, 81, 83–86, 88, 90, 92, 94, 96

**S**
Sustainable calculation method   13, 14
Sustainable construction   9
Sustainable standards   9, 13, 15

**U**
Urban renovation   1, 2, 4, 9–13